AF715145

BLACK HOLES BROKEN DOWN IN AN UNDERSTANDABLE WAY

THE SCIENCE BEHIND THE GRAVITY OF EVERYTHING

JAMES SMITH

Copyright © 2026 by James Smith

All rights reserved.

No part of this book may be reproduced in any form or by any electronic or mechanical means, including information storage and retrieval systems, without written permission from the author, except for the use of brief quotations in a book review.

INTRODUCTION – INTO THE UNKNOWN

There's something about black holes that gets under our skin. They're part mystery, part danger, and part beauty. Even people who don't follow science can't help but feel a shiver of curiosity when they hear those two words. A "black hole" sounds like the ultimate unknown—a place where everything disappears. It's not hard to see why they've become one of the most talked-about ideas in all of astronomy. They combine the things we fear most—darkness, power, and the loss of control—with the things we're most drawn to—mystery, discovery, and the promise of answers hidden in the dark.

Every era has its symbol of mystery. Centuries ago, it was the ocean, vast and unexplored, filled with maps that ended in "Here Be Dragons." Later, it was the unexplored edges of the Earth, then the Moon. Today, it's the black hole—our modern version of the dragon at the edge of the map. Even with all our satellites, telescopes, and theories, we can't look directly into one. That alone sparks the imagination. It's a reminder that no matter how advanced we become, there are still things out there that can't be touched, measured, or neatly explained.

There's also a psychological pull to the idea. Human beings are fascinated by limits—the edge of a cliff, the bottom of the ocean, the boundaries of life and death. A black hole is the ultimate limit written across the universe. It's where time slows, where light vanishes, and where space curves into itself. We're drawn to it for the same reason people watch storms from their windows or stand close to a campfire: it's a safe way to stare into something powerful and dangerous. It gives us a thrill while letting us stay at a comfortable distance.

But black holes aren't just symbols of destruction. They're also places of transformation. Everything that falls toward one—gas, dust, even stars—gets reshaped by it. Around their edges, matter glows brighter than anywhere else in the universe. They're engines of change, not just cosmic trash cans. That mix of darkness and creation makes them more than just scientific curiosities; they're metaphors that reach deep into the way we think about existence itself. They show up in art, poetry, movies, and music not because everyone understands the physics, but because everyone understands the feeling of being pulled toward something vast and unknown.

There's a reason science fiction loves them. Black holes make perfect story devices because they represent the tension between what's known and what might be possible. A writer can send a spaceship near one and suddenly the rules change—time can slow, space can twist, and characters can face questions about destiny, memory, or identity. They allow science and philosophy to meet in the same sentence. Even when they're used loosely or inaccurately, they work because they stand for a universal idea: the edge of understanding.

The fascination isn't limited to storytellers. Scientists themselves have long been captivated by the challenge black holes present. They push mathematics to its breaking point. They force physicists to question whether the universe is as continuous and predictable as it seems. Every time researchers think they've pinned down how gravity or time works, black holes throw a wrench into the model. That's exciting—it's a rare moment when nature tells us, "You haven't got it all figured out yet." For people who spend their lives searching for truth, that kind of mystery is irresistible.

Even their discovery stories have a sense of drama. Early astronomers had only indirect clues—stars moving in strange orbits, light bending in odd ways. They were like detectives piecing together a crime scene without ever seeing the suspect. When the first image of a black hole's shadow was finally captured, it wasn't just a scientific milestone; it was a moment of human triumph. People around the world felt a spark of awe, not because they fully grasped what they were seeing, but because they understood that it meant we had peered into the heart of something once thought unreachable.

On a deeper level, black holes speak to our desire for meaning in a universe that can feel indifferent. If everything can be pulled into a darkness where time and matter dissolve, what does that say about permanence, about life itself? These aren't just scientific questions—they're philosophical ones. The universe gives us an object that embodies disappearance, and yet it exists as part of a larger cycle. Stars die to create them, galaxies are shaped by them, and in some way, the universe uses them to keep renewing itself. That rhythm between ending and renewal reflects something deeply human.

There's also the sheer beauty of the idea. Astronomers describe the glowing disks of matter swirling around black holes—bright rings of energy that blaze before crossing the point of no return. It's a kind of cosmic art, painted in light and gravity. Knowing that something invisible can shape such beauty reminds us that the unseen often has the greatest influence. That message echoes beyond science. The most powerful forces in life—love, fear, curiosity, hope—can't be seen directly either, yet they shape everything we do.

Curiosity itself is part of the story. Every generation of humans has faced something they didn't understand and tried to uncover it. Fire, thunder, disease, the stars—each began as a mystery that frightened us. Once we explored it, we grew. Black holes are just the latest chapter in that long tradition of facing the unknown. They remind us that fear and curiosity often travel together, and that the willingness to ask questions is what moves us forward.

Even the language surrounding black holes captures people's attention. Words like "singularity," "event horizon," and "spaghettification" sound more like poetry than physics. They turn science into something that feels alive, full of texture and rhythm. That helps make an abstract topic feel close to us. You don't need to understand equations to feel the weight of a word like "collapse" or the mystery behind "horizon." It gives people a way to connect emotionally with something that might otherwise seem distant.

Black holes also invite us to rethink our place in the universe. They show that size and strength aren't what define importance. A black hole can be billions of times heavier than our Sun, yet the entire structure fits into a region smaller than a solar system. That contrast between mass and scale challenges our sense of proportion. It's humbling to realize that something so compact can influence galaxies. It puts our own lives into perspective—tiny, yes, but still part of a grand and interconnected story.

Then there's the time factor. The idea that time itself behaves differently near a black hole is one of the most mind-bending concepts in physics. It stretches the imagination to think that an astronaut could approach a black hole and see the universe age thousands of years while only minutes pass for them. That thought alone has inspired endless discussion, art, and even personal reflection. It touches on something ancient: the wish to step outside of time, to see eternity without being crushed by it.

FROM SCIENCE FICTION MONSTERS TO REAL COSMIC OBJECTS

Before astronomers proved they were real, black holes were creatures of imagination—ghostly concepts born in the pages of science fiction and whispered about in physics classrooms. They were often painted as cosmic traps, bottomless pits that devoured everything unlucky enough to wander too close. Movies and novels gave them personalities, turning them into villains of space: giant mouths waiting in the dark, swallowing whole worlds. For decades, they belonged more to storytelling than science. And yet, behind the fiction, there were a few daring thinkers who quietly wondered if these monsters might actually exist.

Long before the term "black hole" was even invented, the idea had already started taking shape in the minds of early scientists. In the 1700s, long before telescopes were powerful enough to detect much beyond our own galaxy, a few physicists speculated about stars so massive that not even light could escape their gravity. They didn't have the math to prove it, but they sensed something possible and unsettling—objects that would be invisible, yet enormously powerful. The thought was quickly tucked away, like an idea too strange to fit the science of its time. It would take centuries before those half-formed guesses began to look less like fantasy and more like fact.

By the early 20th century, physics was changing fast. Albert Einstein introduced his theory of general relativity, showing that gravity wasn't a force that pulled on objects in space—it was space itself bending around mass. That new view of gravity opened the door to ideas no one had considered seriously before. Suddenly, the universe wasn't just a vast clockwork of stars and planets—it was flexible, dynamic, and capable of extremes. Soon after, a German physicist named Karl Schwarzschild found a mathematical solution to Einstein's equations that described a point in space where gravity became infinite. Einstein himself thought it was just a mathematical oddity, not something that could really exist. Yet Schwarzschild's equations quietly sat there, hinting that nature might have stranger things in store than anyone wanted to admit.

When scientists first began discussing these objects, they didn't call them black holes. The early names were awkward—"frozen stars," "collapsed objects," even "singularities." None of them caught on because none of them captured the eerie simplicity of what was being described. In the mid-20th century, American physicist John Wheeler popularized the term "black hole," and with that, a scientific concept became a cultural phenomenon. The name had weight. It carried mystery and danger. It was easy to picture and hard to forget. From that moment, black holes left the realm of dry equations and entered popular imagination with a new identity—half science, half myth.

Science fiction writers were quick to seize the idea. To them, black holes were more than physical objects—they were perfect symbols. They could represent death, transformation, or the terrifying pull of the unknown. In one story, a ship might be torn apart by a black hole's gravity; in another, it might use the hole as a doorway to another universe. They became backdrops for stories about survival, time travel, and the limits of human courage. Even though the science was often exaggerated or misunderstood, those stories kept public interest alive long before real evidence appeared. People might not have understood the math, but they understood the feeling: the mix of awe and fear that comes with facing something far bigger than ourselves.

The truth, as it turned out, was both more ordinary and more extraordinary than fiction. As technology advanced, astronomers began finding signs that black holes weren't just theoretical—they were out there, shaping galaxies. The "monsters" from stories turned out to be natural consequences of how stars lived and died. When massive stars exhausted their fuel, they collapsed under their own gravity, leaving behind something so dense that even light couldn't escape. That simple, brutal process—death by gravity—created real black holes scattered across the universe. The more scientists looked, the more they found. What had once been science fiction began showing up in real data, real images, and real equations.

The transition from myth to reality didn't happen overnight. Early observations were full of skepticism. Some scientists argued that the strange behaviors seen in deep space—stars orbiting invisible partners, radiation pouring from unseen regions—must have other explanations. But each new piece of evidence fit too neatly to ignore. By the time researchers confirmed the presence of supermassive black holes at the centers of galaxies, the case was undeniable. The cosmic predators of old stories had turned out to be real, though not in the way people once feared. They weren't wandering threats waiting to devour planets; they were key parts of how the universe organized itself.

That realization changed how scientists—and the public—talked about them. The fear gave way to fascination. Instead of thinking of black holes as destructive forces, astronomers began to see them as engines of creation. They helped galaxies form. They influenced the movement of stars. Their gravity sculpted the structure of space on enormous scales. In a sense, the "monster" had become the architect. It was a shift that mirrored how humans often move from superstition to understanding: what begins as fear of the unknown eventually becomes respect for the complexity behind it.

Even now, the mix of myth and reality still lingers. Hollywood continues to use black holes as symbols of danger or mystery, while scientists quietly work out their details through observation and data. The famous photograph of the black hole in galaxy M87, released in 2019, was one of those rare moments when the two worlds briefly touched. People who had grown up watching fictional depictions of black holes suddenly saw a real one for the first time. The blurry orange ring wasn't the swirling tunnel of a movie, but it was far more meaningful—it was proof. It showed that human curiosity could reach across unimaginable distances and bring back something real.

The shift from fiction to fact didn't make black holes less mysterious; it made them more profound. Knowing that they exist doesn't drain away their wonder—it deepens it.

They remind us that the universe doesn't need to invent monsters for us. Nature itself is capable of producing things more extreme and awe-inspiring than anything a writer could dream up. Yet, at the same time, they teach a subtle lesson: reality and imagination aren't opposites. One feeds the other. The stories we tell about the universe inspire us to explore it, and the discoveries we make feed new stories in return.

HOW STUDYING BLACK HOLES HELPS US UNDERSTAND THE UNIVERSE ITSELF

Every discovery about black holes brings us closer to answering one of humanity's oldest questions: what is the universe made of, and how does it work? Black holes may look like the ultimate dead ends—points where everything stops—but in truth, they act like windows into the deepest workings of nature. They test the rules of physics in ways nothing else can. The closer scientists look at them, the more they reveal about how the cosmos forms, evolves, and holds itself together.

At first, that might seem strange. How could something we can't see directly teach us so much? The answer lies in the way black holes affect everything around them. They bend light, twist space, slow time, and release enormous amounts of energy. By studying these effects, scientists can trace the outlines of invisible forces and hidden structures that shape the universe on both the smallest and largest scales. Black holes are like stress tests for reality—they show what happens when the normal rules are pushed to their breaking point.

One of the biggest lessons black holes teach us is about gravity itself. Einstein's general theory of relativity describes how gravity isn't a pull between objects, as Newton thought, but rather a bending of space and time caused by mass. The more mass something has, the more it warps the fabric of space around it. Black holes take that idea to the extreme. Their gravity is so concentrated that it creates a region from which nothing can escape. When scientists study how light behaves near these regions—how it bends, stretches, or disappears—they can test whether Einstein's equations still hold up. Every successful prediction strengthens our confidence in relativity, and every unexpected result points to something new waiting to be understood.

But gravity isn't the only part of the story. Black holes also help explain how galaxies themselves come to be. At the center of almost every large galaxy—including our own Milky Way—sits a supermassive black hole millions or even billions of times heavier than the Sun. These giants seem to act as anchors for the galaxies around them. Their immense gravity helps pull matter together, shaping how stars form and where they cluster. In the early universe, when everything was still chaotic, black holes may have played a key role in turning scattered clouds of gas into organized galaxies. Without them, the cosmos might look far less structured than it does today.

The connection between black holes and galaxies works both ways. As gas and dust fall toward a black hole, they don't simply vanish—they heat up and emit intense radiation. That energy can blow out enormous winds that affect entire galaxies, slowing or stopping the formation of new stars. In this way, a black hole can act like a cosmic regulator, keeping its host galaxy in balance. It's a delicate relationship: the galaxy feeds the black hole, and the black hole, in turn, shapes the galaxy's growth. Understanding that partnership helps scientists explain why galaxies come in so many different sizes and forms.

Black holes also offer insight into the nature of light and energy. When matter spirals toward one, it forms an accretion disk—a bright, swirling ring of superheated particles moving at nearly the speed of light. This process converts gravitational energy into radiation more efficiently than any other known mechanism in the universe. Some black holes even produce powerful jets that shoot out from their poles, stretching for thousands of light-years. These jets pump energy back into the surrounding space, influencing how nearby matter behaves. By studying them, astronomers learn not only about the black holes themselves but also about how energy moves through the cosmos.

The information black holes provide doesn't stop at the visible universe. They also hold clues about the invisible parts—the dark matter and dark energy that make up most of what exists. While black holes aren't made of dark matter, their gravitational effects help scientists test how matter behaves on cosmic scales. Observing how stars and gas move around black holes allows researchers to refine models of how gravity operates in extreme conditions. Those same models can then be used to explore mysteries like why the universe's expansion is accelerating or how unseen mass shapes galaxies.

Another major area where black holes have changed our understanding is time. Near a black hole, time doesn't flow the way we experience it on Earth. The stronger the gravity, the slower time moves. Astronauts traveling near a black hole would experience minutes while centuries pass elsewhere. This isn't just science fiction—it's a measurable effect of relativity. By studying these time distortions, scientists can refine our understanding of how time and space are linked. It also helps test technologies like GPS satellites, which must account for small relativistic effects to stay accurate. In that sense, the physics of black holes reaches all the way down to the devices we use every day.

Even the smallest black holes can reveal something profound. When two of them collide, they send ripples through the fabric of space-time—gravitational waves that travel across the universe. These waves were first detected in 2015, confirming another one of Einstein's predictions. Every new detection adds more detail to our map of the universe. It's like hearing the cosmos for the first time. These waves carry information about events billions of light-years away, giving us a way to study regions that light can't reach. Through them, we can observe the birth of black holes, the merging of neutron stars, and even glimpses of what the early universe was like right after the Big Bang.

Black holes also force us to confront the limits of our understanding. At their center lies something called a singularity, a point where density becomes infinite and our laws of physics break down. Right now, no theory can describe what happens inside that boundary. General relativity can explain large-scale gravity, and quantum mechanics can explain the behavior of tiny particles—but the two don't fit together inside a black hole. That conflict has become one of the biggest challenges in modern physics. By trying to understand what happens at a black hole's core, scientists are working toward a unified theory of everything—a framework that could explain both the vastness of space and the smallest pieces of matter in one elegant picture.

That search has already led to remarkable ideas. Physicist Stephen Hawking proposed that black holes aren't entirely black. They slowly release tiny amounts of radiation, meaning they can shrink and eventually vanish over unimaginable timescales. This discovery linked gravity, quantum theory, and thermodynamics in a way no one had expected. It suggested that black holes might store and release information in subtle ways, hinting at deep connections between energy, entropy, and the fundamental nature of reality. What began as an effort to describe the darkest places in the universe ended up reshaping how scientists think about the laws of nature themselves.

1

WHAT EXACTLY IS A BLACK HOLE?

To understand what makes a black hole so extraordinary, it helps to start with something completely ordinary: gravity. It's a force we live with every moment of our lives. It's what keeps our feet on the ground, what pulls rain toward the Earth, and what keeps the Moon circling our planet. But gravity isn't a force that we can see—it's a relationship between objects with mass. The more massive something is, the stronger its pull. The farther away you are from it, the weaker that pull becomes. It's such a simple idea that we take it for granted. Yet when taken to the extreme, gravity can do things that defy our everyday understanding.

A black hole is the ultimate example of gravity pushed to its limits. It's a region of space where gravity is so intense that nothing—not even light, the fastest thing in existence—can escape its pull. That's what makes it "black." Light doesn't bounce off its surface or travel away into space; it's trapped, curved back in on itself by the sheer warping of space and time. Once something crosses a certain boundary, known as the *event horizon*, it's cut off from the rest of the universe forever.

This idea might sound abstract, but the principle behind it is surprisingly straightforward. Every planet, moon, and star has an escape velocity—the speed you'd need to move in order to break free from its gravitational pull. On Earth, that speed is about seven miles per second. That's why rockets have to reach such high velocities to leave the atmosphere. The Moon's gravity is weaker, so the escape velocity there is lower. But if you were standing on an object much more massive and compact than Earth—say, a neutron star—you'd need to travel at an incredible fraction of the speed of light just to get away. A black hole takes this one step further: its escape velocity is faster than light itself. And because nothing can move faster than light, nothing can escape.

It's not that light is "sucked in" or destroyed. Light follows the natural curves of space, and near a black hole, those curves are so extreme that every possible path leads inward. Space and time aren't separate things—they're woven together into a single fabric called *space-time*. When an object with mass sits in space-time, it bends that fabric around it. The more massive the object, the deeper the bend. A black hole is a point where the bend becomes infinite—so deep that it creates a one-way path inward.

One way to think about this is to picture space as something that can stretch and curve. On Earth, gravity feels like a downward pull, but what's really happening is that space-time itself is curved around our planet, guiding objects along its surface. When you drop an apple, it's not being pulled down by an invisible string—it's following a curved path in space-time that leads to the center of the Earth. Now imagine an object so dense that it curves space completely inward, forming a pit that nothing can climb out of. That's a black hole.

At the edge of that pit lies the event horizon, the boundary that separates what can escape from what can't. You could hover near it—if you had some impossible source of propulsion—and see light bending and distorting around you. But once you cross that invisible line, every possible path forward leads deeper inside. Even if you fired your engines at full power or shot a beam of light straight upward, space-time itself would carry it back down. The escape routes don't exist.

The concept of something with gravity so strong that light couldn't escape was actually proposed long before black holes were officially discovered. In the late 1700s, a few scientists speculated about "dark stars," massive objects whose light couldn't reach us. They didn't have Einstein's theory of relativity to describe how space and time bend, but the intuition was correct. It took more than a century—and a lot of skepticism—before the idea reemerged with the math to back it up. When Einstein's equations showed that space could curve infinitely under enough mass, the pieces began to fit.

The key word here is *density*—how much mass is packed into a given space. You could take something the size of the Sun and compress it into a sphere only a few kilometers across, and it would become a black hole. It's not the total mass alone that matters, but how tightly it's squeezed together. Even a mountain could become a black hole if it were crushed into an unimaginably small volume. The idea is shocking because it forces us to confront how flexible the universe really is. Matter and space aren't fixed things—they can stretch, shrink, and fold into new forms.

What makes black holes so fascinating is that they reveal

how gravity isn't just a force—it's a shape. Around Earth, space curves gently. Around a black hole, it folds in on itself. This curvature affects everything, including time. Near a black hole, time slows dramatically. To an outside observer, anything approaching the event horizon appears to move slower and slower, fading into stillness just before crossing the boundary. To the object itself, though, time feels normal —it passes right through, unaware that the rest of the universe has rushed far ahead. That distortion of time is one of the strangest but most well-tested predictions of relativity.

Even though light can't escape a black hole, the region around it can be incredibly bright. As matter spirals inward, it forms a disk of superheated gas and dust that glows fiercely in X-rays and visible light. The closer the material gets to the event horizon, the faster it spins and the hotter it becomes, releasing more energy than almost any other process in the universe. This is how scientists can detect black holes—by watching the light and motion of the matter around them, even if the black hole itself remains invisible.

The boundary where light is trapped isn't a physical surface; it's a point where space-time becomes so warped that all directions lead inward. Beyond that, the concept of "inside" and "outside" loses meaning. It's not that the black hole is an object sitting in space—it's that space itself has been rearranged around it. Every bit of mass, every ray of light, every second of time that crosses that horizon becomes part of the same geometry.

This is why black holes continue to puzzle even the most advanced physicists. They represent the place where our best theories stop agreeing with each other. General relativity perfectly describes the warping of space-time, but it can't handle the infinite density of the singularity inside.

Quantum physics, which rules the world of atoms and particles, breaks down under that gravity. Somewhere between those two extremes lies a deeper theory—something that explains both gravity and quantum behavior in the same language. Black holes might be the key to finding it.

Despite all their complexity, the defining feature of a black hole comes back to that simple principle: it's a region where gravity overwhelms everything else. No amount of energy, no beam of light, no force we know can overcome it once it crosses the boundary. It's not malevolent or mysterious in the supernatural sense—it's just nature following its own laws to their absolute limit.

GRAVITY AND ESCAPE VELOCITY

Gravity is one of those forces that feels familiar but is hard to describe precisely. It's invisible, constant, and inescapable. Every moment of your life, it's at work—keeping your body on the ground, guiding the tides, and shaping the motion of planets and stars. We don't notice it most of the time because it's always there, quietly balancing everything. Yet this simple, everyday force is the reason black holes exist. To understand what makes them so powerful, it helps to first understand what gravity really does and how it relates to the idea of escape velocity.

At its heart, gravity is the attraction between anything that has mass. You have mass. So does your chair, the Earth beneath you, and the Sun above. Every bit of matter pulls on every other bit, no matter how far apart they are. Most of the time, these pulls are so small we don't notice them. You're technically attracting your coffee mug right now, but the Earth's mass is so much greater that its pull overwhelms every smaller force around you. Gravity is proportional to mass—the bigger the mass, the stronger the attraction.

It also depends on distance. The farther away two objects are, the weaker the gravitational pull between them becomes. Double the distance, and the pull drops to one quarter of its strength. That's why the Moon doesn't crash into Earth, and why you don't feel pulled toward the Sun more strongly than the ground beneath your feet. The Sun's gravity keeps our planet in orbit, but the distance between them keeps that pull perfectly balanced with our planet's motion.

What's fascinating is that gravity doesn't just affect objects—it affects space itself. Einstein's theory of general relativity describes gravity not as a force in the traditional sense, but as a bending of space and time caused by mass. A planet curves the space around it, and that curvature guides the paths of other objects. The Moon doesn't circle Earth because it's being tugged by an invisible string; it's simply following the curved path that Earth's mass creates. This bending of space is subtle on small scales, but when you're dealing with stars, planets, or black holes, it becomes dramatic.

That curvature explains why things fall. When you drop a ball, it's not being "pulled" downward—it's following the natural curve of space-time toward the center of Earth's mass. In that sense, gravity isn't something acting on an object from afar. It's the shape of space telling the object how to move. The deeper the curvature, the harder it is for something to move away. That idea connects directly to escape velocity.

Escape velocity is the speed an object needs to overcome a planet's—or any body's—gravitational pull. If you toss a ball upward, gravity slows it down until it falls back. Throw it harder, and it goes higher before returning. But if you could throw it fast enough, it would never come back at all. That threshold speed is the escape velocity. It's the tipping point where motion outward wins over gravity pulling inward.

On Earth, the escape velocity is about seven miles per second, or roughly 25,000 miles per hour. That's why rockets need such enormous energy—they have to push past that limit to break free of the planet's gravity. The more massive a planet or star is, and the smaller its size, the higher its escape velocity becomes. If Earth were smaller but kept the same mass, its gravity would be stronger at the surface, and the speed needed to escape would rise. That relationship between mass, size, and gravity is what makes black holes so extreme.

But before getting to that level, it's helpful to see how this concept works across scales we can relate to. Take the Moon, for instance. It's about one-sixth the mass of Earth, and its escape velocity is just over 2 miles per second. That's why astronauts on the Apollo missions didn't need as much fuel to leave its surface as they did to leave Earth's. The lower gravity meant an easier escape. On the other hand, Jupiter, the largest planet in our solar system, has an escape velocity of about 37 miles per second. You'd need much more power to break free from its gravitational pull because its mass—and therefore its curvature of space—is so much greater.

Now stretch that concept further. If you kept compressing matter into smaller and smaller spaces, the escape velocity would rise dramatically. There's a point where the escape velocity becomes the speed of light itself —about 186,000 miles per second. Once that happens, no light, no particle, no signal of any kind can get out. That's the condition that defines a black hole. The curvature of space around it becomes so steep that every possible path bends back inward.

It's a simple idea with profound consequences. If you could somehow stand near the edge of a black hole, known as the event horizon, you'd find that no matter how fast you tried to move away, gravity would always win. Even beams of light—normally the universe's ultimate speed limit—can't escape. That's why the black hole appears black. It's not that it destroys light, but that light can't find a route outward. Space itself is shaped in such a way that all roads lead back to the center.

Understanding escape velocity also helps explain why black holes don't behave like cosmic vacuum cleaners. Gravity doesn't grow stronger just because something becomes a black hole; it's only extreme close to its surface. Farther away, the pull weakens according to the same rules as any other object. If the Sun were somehow replaced with a black hole of equal mass, Earth would continue orbiting as it always has—dark and frozen, but still in the same path. The key difference is that near a black hole's surface, gravity is concentrated into such a small region that the escape velocity exceeds what physics allows.

Another interesting aspect is that gravity doesn't act instantly. Changes in gravity—like when objects move or collapse—travel at the speed of light. That means when a star collapses into a black hole, the gravitational influence doesn't suddenly change throughout the universe. It ripples outward, adjusting the fabric of space as it goes. This behavior is what allows scientists to detect gravitational waves, those faint ripples in space-time created by massive events like the merging of two black holes. Each wave carries information about how gravity behaves under the most extreme conditions imaginable.

The concept of escape velocity might sound mathematical, but it's deeply connected to intuition. You can feel it whenever you throw something upward or watch a rocket launch. Every motion upward is a struggle against gravity's hold. For most of us, that struggle ends quickly—a ball returns to your hand, a jumper lands back on the ground. But scale that up, and you get the story of how worlds hold onto atmospheres, how moons stay in orbit, and how stars live and die. It's the same principle, applied across the cosmos.

WHY THEY'RE "BLACK" AND WHY WE CAN'T SEE THEM DIRECTLY

The name *black hole* sounds dramatic, but it's more literal than poetic. They're called "black" because they emit no light, and "hole" because they act as a kind of one-way region in space—things can fall in, but nothing can come back out. Yet the real reason we can't see them has nothing to do with them being objects that absorb light, like charcoal or asphalt. Instead, it's because light itself cannot escape from within their boundaries. The space around them is bent so sharply by gravity that every possible path light could take loops back inward. From our perspective, it's as if they've vanished from the visible universe.

Everything we see—stars, planets, dust clouds—depends on light reaching our eyes or our telescopes. Light is our messenger. Whether it's visible, infrared, or radio, all of astronomy relies on detecting photons that have traveled across vast distances to bring us information. A black hole interrupts that process. It's the ultimate silence in a universe full of light. When light falls into it, it doesn't reflect, scatter, or bounce—it disappears from view. That's what makes the region truly "black."

The darkness isn't just an absence of light—it's the result of how space and time behave near a black hole. Gravity doesn't just pull on matter; it also bends the path that light takes. On Earth, the effect is tiny. Light travels in straight lines because our planet's gravity isn't strong enough to bend it noticeably. But near a black hole, gravity is so extreme that it changes the very geometry of space. Light rays curve around it like water circling a drain. If they cross the event horizon—the invisible boundary where the escape velocity equals the speed of light—they can't climb back out. Even light traveling straight outward finds itself dragged back in.

From a distance, this creates a strange optical illusion. Around a black hole, there's often a glowing halo called the *accretion disk*—hot gas and dust swirling at incredible speeds just outside the event horizon. That disk shines brightly, sometimes outshining entire galaxies. But the center remains perfectly dark. The contrast between that luminous whirlpool and the void at its core gives black holes their iconic appearance. It's not that the hole itself emits darkness; it's that no light escapes to replace what's lost.

The closest analogy is a perfectly black surface—a material that absorbs all light and reflects none. But even that doesn't come close. A true black hole doesn't absorb light in the same way a physical object does; it bends light out of existence as far as outside observers are concerned. If you were to shine a flashlight directly toward a black hole, you wouldn't see a beam bouncing back. The photons from that beam would follow the curvature of space until they vanished past the event horizon. It's not that they were destroyed; they're simply trapped in a region from which information can never return.

That's also why we can't "see" black holes directly. Telescopes depend on collecting light, but a black hole emits none of its own and reflects none from elsewhere. Instead, astronomers must detect them indirectly, by studying the effects they have on their surroundings. When a black hole pulls on nearby gas or stars, we can measure the movement of those objects. If a star orbits something invisible yet incredibly massive, we can infer that a black hole is there. In that way, it's less like spotting an object and more like noticing the shadow it casts on space itself.

Even though the black hole itself is invisible, the region around it can be spectacularly bright. As gas spirals inward, it heats up from friction and compression, reaching millions of degrees. That heat produces radiation—from X-rays to visible light—that we can detect. It's the light from the edge, not the center, that reveals a black hole's presence. Think of it like a campfire surrounded by darkness: you can't see the blackness itself, but the glow at its edge tells you something powerful is there.

The reason that light behaves this way lies in the relationship between gravity and geometry. In ordinary space, a beam of light moves in straight lines. But a black hole bends those lines until they curve back on themselves. The closer the light gets to the event horizon, the more tightly it bends. Just outside that edge is a zone called the *photon sphere*, where light can orbit the black hole in circles. A photon traveling through this region might make several loops before either escaping outward or being pulled in completely. That's why the light around black holes often appears distorted and magnified—some of it comes from behind the black hole, bent around by gravity in a kind of cosmic lensing effect.

When astronomers first captured an image of a black hole's shadow, they weren't seeing the black hole itself but this interplay of light around it. The photograph, released in 2019, showed a glowing ring surrounding a dark center. That dark circle wasn't the event horizon; it was the shadow cast by it, created by the way light bends around the black hole. It was humanity's first visual confirmation that these invisible objects truly exist.

There's also a deeper reason the word "black" fits beyond its visual description. In physics, black holes represent boundaries—limits to what can be known. Once something crosses the event horizon, no signal, no particle, not even information can make its way back to us. It's as if the universe draws a curtain, hiding whatever happens inside. That mystery gives black holes their eerie reputation. They aren't just invisible in the literal sense; they're also unknowable beyond a certain point. The laws of physics as we understand them stop functioning clearly inside. Space, time, and matter blend into something we can't yet describe.

Despite that mystery, we've found ways to trace their outlines. By studying how stars move around invisible centers, or how matter heats up before falling in, astronomers can map the hidden presence of black holes across the cosmos. They've been detected in the hearts of galaxies, in binary star systems, and even through the gravitational waves they produce when they merge. Each detection tells us that the "blackness" isn't emptiness—it's a boundary where reality curves too sharply for light to climb out.

It's easy to think of black holes as places where things disappear, but the more accurate picture is that they're places where everything—matter, light, energy—reaches a point of no return. The energy doesn't vanish; it's simply confined beyond reach. From the outside, all that energy contributes to the black hole's mass, increasing its pull and widening its event horizon. Over time, black holes can grow by feeding on nearby matter, but their blackness never changes. No matter how massive they become, they never glow from within.

That idea—total invisibility paired with immense influence—makes black holes one of the most paradoxical features of the universe. They are both absent and present. You can't see them, yet they shape the motion of entire galaxies. They leave no reflection, yet they light up their surroundings with the energy they release indirectly. They're reminders that darkness isn't always the opposite of light—it can be the condition that defines it.

2

HOW BLACK HOLES ARE BORN

Every star in the sky, from the faintest flicker to the brightest beacon, has a beginning, a middle, and an end. They may seem timeless, but stars are born, grow old, and eventually die—just on a scale so immense that we only see snapshots of their lives. Understanding the life cycle of a star helps us see how black holes come to exist, because every black hole begins as a star that's run out of time. Yet before that final transformation, a star lives a long, dynamic life filled with constant change and incredible energy.

It all starts in a cloud—a vast, cold region of gas and dust floating between the stars. These clouds are called nebulae, and they're like the nurseries of the universe. Inside them, hydrogen, helium, and tiny particles drift quietly until gravity begins to pull things together. Over time, clumps form, growing heavier and denser as more material gathers. As gravity strengthens, these clumps begin to shrink, and the pressure inside them rises. When that pressure becomes great enough, the temperature soars to millions of degrees.

At that moment, the core becomes hot enough for nuclear fusion to begin. That's the instant a star is born.

Fusion is what defines a star. It's the process that turns hydrogen—the simplest element—into helium, releasing enormous amounts of energy in the process. This energy pushes outward, balancing the inward pull of gravity. That balance creates stability, allowing a star to shine steadily for millions or billions of years. The light that reaches Earth from our Sun is the result of countless fusion reactions happening every second in its core. Each reaction converts a small amount of mass into pure energy, following Einstein's famous equation, $E=mc^2$. It's an incredibly efficient engine, and once it starts running, it can keep going for eons.

The early years of a star's life are relatively peaceful. Once it settles into balance, it spends most of its lifetime in what astronomers call the *main sequence* stage. During this period, the star's core fuses hydrogen into helium at a steady rate, maintaining that delicate tug-of-war between gravity and pressure. The star's size and temperature determine how long it will stay in this phase. Smaller stars burn their fuel slowly, lasting tens or even hundreds of billions of years. Larger stars burn through their fuel much faster, shining brighter but living shorter lives. The biggest stars in the universe can exhaust their hydrogen in just a few million years—a blink of an eye in cosmic time.

The Sun, for example, is about halfway through its main sequence life. It's been shining for roughly 4.6 billion years and has about that much time left before major changes begin. It's a stable, middle-aged star, neither too young nor too old, maintaining a steady glow that sustains life on Earth. But nothing in the universe stays the same forever. Eventually, the supply of hydrogen in the core runs low.

When that happens, the balance that keeps the star stable starts to falter.

As hydrogen runs out, the fusion reactions slow, and gravity begins to win. The core starts to contract, becoming denser and hotter. This rising temperature reignites fusion in the outer layers, causing the star's outer shell to expand dramatically. The star swells into what's known as a red giant. Its surface cools as it grows, turning redder, but its size can balloon to hundreds of times its original diameter. If you replaced our Sun with a red giant of the same mass, its outer layers would engulf Mercury and Venus, and might even reach Earth's orbit.

Inside, the story is more complex. The shrinking core heats up enough for helium atoms to fuse into heavier elements like carbon and oxygen. For a while, that gives the star a second wind. But this phase is unstable and temporary. When the helium runs out, the core contracts again, and the star faces its next transformation. What happens next depends entirely on how massive the star is.

For smaller and medium-sized stars like the Sun, the end is relatively calm. The outer layers drift away into space, forming a glowing shell of gas called a planetary nebula. The remaining core, about the size of Earth but incredibly dense, becomes a white dwarf. This remnant no longer burns fuel; it simply glows with the leftover heat of its past life. Over billions of years, it will slowly cool and fade into darkness. That's the quiet retirement of a typical star.

For massive stars, the ending is anything but quiet. When their cores can no longer support the weight of the outer layers, the collapse is sudden and violent. The force of gravity crushes the core, triggering a supernova—a colossal explosion that blasts the star's outer layers into space. For a brief moment, the explosion can shine brighter than an entire galaxy. The light and energy released scatter elements like iron, gold, and oxygen across space, seeding future generations of stars and planets. Without those explosions, the universe would be a much simpler place, lacking the elements needed for complex worlds and life itself.

After the supernova fades, what remains depends on the size of the star's original core. If the core was massive enough, the collapse continues beyond any known limit. The matter compresses to the point where even atomic particles can't resist the pull of gravity. At that stage, a neutron star or a black hole is born. Neutron stars are incredibly dense—just a few miles across, yet containing more mass than the Sun. But when the mass is greater still, not even neutrons can withstand the pressure. The core shrinks past the point of no return, and a black hole forms.

The life and death of a star are deeply connected to everything around it. Stars are the universe's factories, creating the elements that make planets, oceans, and living things possible. Every atom of calcium in your bones and iron in your blood was once forged inside a star that lived and died long before the Sun existed. When stars end their lives in explosions, they release those elements into space, where they can become part of new stars and planets. The cycle repeats endlessly, building the universe piece by piece.

WHAT HAPPENS WHEN MASSIVE STARS RUN OUT OF FUEL

The lives of massive stars end not with a quiet fade, but with a spectacular finale. These stars, many times larger than our Sun, live fast and burn bright. Their size gives them immense power, but it also seals their fate. The same forces that make them shine so intensely also lead to their destruction. When they run out of fuel, gravity takes over completely—and what happens next reshapes entire regions of the universe.

During their lifetimes, massive stars hold themselves together through a delicate balance. Deep in their cores, nuclear fusion turns hydrogen into helium, releasing huge amounts of energy. That energy pushes outward, creating pressure that keeps the star from collapsing under its own weight. The larger the star, the stronger the gravity trying to pull it inward, and the more energy it must generate to stay stable. But fusion depends on having fuel, and the supply is not endless.

Over millions of years, the star's hydrogen begins to run low. As the hydrogen burns away, the core becomes dominated by helium. The fusion process shifts, and the star must adapt. The core contracts under gravity, becoming hotter and denser. The increased temperature allows new types of fusion to begin—helium fusing into carbon and oxygen. The outer layers expand, and the star grows into a red or blue supergiant, depending on its size and temperature. Its surface may cool, but inside, the core is a raging furnace, cycling through new stages of fusion as heavier elements form.

Each stage of fusion burns faster than the one before it. Hydrogen can sustain a star for millions of years, but helium lasts only a fraction of that time. When the star begins fusing carbon, the timeline shortens even more—each new element takes less and less time to consume. By the time the core is creating elements like silicon or iron, the entire process happens in mere days. The star's life accelerates toward an inevitable collapse, like a candle burning down to its wick.

Then it reaches the point where fusion can go no

further. When the core fills with iron, the process stops. Iron marks a dead end for fusion because it can't release energy by fusing into heavier elements. Instead of producing energy, fusing iron actually consumes it. At that moment, the delicate balance that has held the star stable for millions of years breaks. Gravity, which has been held at bay by the pressure from fusion, suddenly wins.

The collapse happens in an instant. The core, no longer supported by outward pressure, implodes under its own weight. It falls inward at incredible speed, and the temperature skyrockets—reaching billions of degrees in fractions of a second. Protons and electrons are crushed together, forming neutrons. The core shrinks from the size of a planet to something only a few miles across, denser than anything we can easily imagine. The outer layers of the star, still racing inward, slam into this dense neutron core and rebound outward in a massive shockwave. That shockwave tears through the star, blasting its outer layers into space.

This event is what we call a supernova—one of the most powerful explosions in the universe. For a short time, a single supernova can outshine an entire galaxy. It releases more energy in a few seconds than the Sun will produce over its entire lifetime. The explosion sends heavy elements —carbon, oxygen, gold, iron, and others—scattering across space. Those elements become the raw material for new stars, planets, and eventually, life.

What's left behind depends on how massive the star was to begin with. If the core's remaining mass is between about one and three times that of the Sun, it becomes a neutron star—a city-sized sphere made entirely of neutrons. It's so dense that a teaspoon of its material would weigh billions of tons. But if the core is heavier than that, not even the pressure of neutrons can stop gravity's collapse. The core continues to shrink, folding space around it until it becomes a black hole.

From the outside, it's easy to think of a star's collapse as destruction, but it's more accurate to think of it as transformation. The energy and matter released in a supernova are not lost—they're redistributed. The explosion seeds the galaxy with elements that will form new solar systems. The core, though it may collapse into something dark and invisible, remains part of that ongoing cosmic cycle.

There's a strange beauty in how nature balances creation and annihilation in these moments. The same fusion that once kept the star alive created the heavy elements that make new worlds possible. And when fusion stops, the resulting collapse ignites new life across the cosmos. Every atom of iron in your blood was once formed in the core of a dying star. The oxygen you breathe, the calcium in your bones, the gold in a wedding ring—all of it was forged in the final seconds of some ancient supernova.

The collapse of a massive star isn't just an end—it's also a doorway into the universe's most extreme physics. During those final milliseconds, matter behaves in ways that defy intuition. Gravity becomes overwhelming, densities reach unimaginable levels, and the structure of space-time itself starts to warp. These are the conditions that give birth to neutron stars and black holes, objects so strange that they challenge our understanding of reality.

Even after the explosion fades, the star's influence lingers. The shockwave from a supernova can compress nearby clouds of gas, triggering the formation of new stars. It's a kind of cosmic recycling, where the death of one star leads to the birth of many others. This process has been happening for billions of years, shaping galaxies and filling them with the elements that make planets and life possible.

The universe is full of these echoes—traces of long-gone stars that lived, burned out, and exploded. Astronomers can still detect the remnants of supernovae as glowing clouds of gas, expanding shells that drift across space for thousands of years. Some of them contain neutron stars at their centers, spinning rapidly and sending out beams of radiation. Others hide black holes, invisible but revealed through their gravitational pull on nearby stars. These remnants are the ghosts of stars, silent but powerful reminders of the forces that built the cosmos.

SUPERNOVA EXPLOSIONS AND THE LEFTOVER CORE

When a massive star reaches the end of its life, the universe doesn't let it slip quietly into darkness. Instead, it ends in an event so powerful it can light up a galaxy—a supernova. This is nature's way of clearing the stage and recycling the materials of a dying star. It's both destruction and creation at once: the star's outer layers are torn apart in a brilliant explosion, while its core collapses into something entirely new.

A supernova begins in the star's center, deep beneath its layers of gas and fusion. Once the core stops producing energy, gravity gains the upper hand. There's no more outward pressure to balance the pull inward, and everything begins to fall. The collapse is shockingly fast, taking only seconds. The core, once the size of a small planet, is crushed down to something no larger than a city. This sudden compression creates an extraordinary rebound effect. The outer layers of the star, still plummeting inward, crash into the now-solid core and bounce outward violently. That rebound, amplified by an enormous surge of energy from trapped neutrinos, rips through the star and blasts its outer shell into space.

The result is an explosion that can release more energy in a few moments than our Sun will produce over its entire lifetime. The brightness of a supernova can outshine billions of stars. To an observer in space, it would look as if a new star had suddenly appeared where none existed before, then gradually faded over weeks or months. But what's left afterward is not a new star—it's the shattered remains of the old one.

The expanding cloud of debris left behind is called a *supernova remnant*. It's a chaotic mix of gas and heavy elements moving outward at thousands of miles per second. That material enriches the surrounding space, seeding it with elements forged inside the star's final moments—iron, gold, oxygen, and more. Those ingredients drift through the galaxy and eventually become part of new stars, planets, and even living organisms. Every element heavier than hydrogen and helium owes its existence to the death throes of stars like these. In a sense, supernovae are the universe's way of spreading the building blocks of everything.

But while the outer layers are blown apart, the core tells a very different story. What happens to that dense center depends on how massive the original star was. If the remaining core is about one and a half to three times the mass of our Sun, it becomes a *neutron star*. These objects are among the strangest in existence. A neutron star packs more mass than the Sun into a sphere barely twelve miles across. The gravity is so strong that the atoms themselves are crushed. Electrons and protons combine to form neutrons, leaving behind matter in its densest possible form short of a black hole.

A neutron star's surface is unimaginably extreme. The

gravity is so intense that if you dropped something from a few feet above, it would hit the surface at tens of millions of miles per hour. A teaspoon of its material would weigh billions of tons. Despite their tiny size, neutron stars can spin hundreds of times per second. Some emit beams of radiation that sweep across space like lighthouse beams; these are called pulsars. When the beam crosses Earth, it appears as a regular pulse of light, precise enough that astronomers can use them as cosmic clocks.

If the core of the collapsing star is even more massive—greater than about three times the mass of the Sun—gravity doesn't stop at a neutron star. It keeps going. No known force can resist the pull. The collapse continues until the core becomes a black hole, a point where matter and space are so compressed that not even light can escape. The event horizon forms, marking the boundary between what can exist in the outside universe and what is lost to the depths of gravity's well. The black hole begins as a remnant of death, but it becomes something new entirely: a structure defined not by matter, but by space itself folding inward.

Supernova explosions are not uniform. Some are more violent than others. The type of supernova depends on the size and composition of the star. Massive stars that still have their hydrogen envelopes explode as what astronomers call *Type II supernovae*, while stars that have lost much of their outer material before collapsing produce other types. In every case, though, the same principle holds: gravity pushes inward, pressure fails, and energy is released in a final, cataclysmic burst.

During those moments, the physics of the universe is pushed to its extremes. Temperatures rise to billions of degrees, and new elements are forged in the chaos. In the brief heart of the explosion, atomic nuclei collide and merge, creating elements heavier than iron—things like nickel, copper, silver, and gold. Without these violent deaths, the universe would be a simpler place, filled only with the lightest elements that formed after the Big Bang. Supernovae are, in a sense, the cosmic furnaces that add richness and complexity to everything that follows.

Even though supernovae mark the end of a star's life, they also give rise to some of the universe's most beautiful sights. The expanding remnants of these explosions glow for thousands of years, shaped by shockwaves and magnetic fields. One famous example, the Crab Nebula, is the aftermath of a supernova observed on Earth nearly a thousand years ago. What remains is a glowing cloud of filaments and gas, with a pulsar—a rapidly spinning neutron star—at its center. It's a reminder that even destruction can create something breathtaking.

The violence of a supernova also influences its surroundings on a larger scale. The explosion releases enough energy to push and compress nearby clouds of gas, triggering new waves of star formation. In this way, the death of one star can spark the birth of many others. Galaxies are filled with these cycles of creation and collapse, linked together by the shockwaves of ancient supernovae. The heavy elements produced in one generation of stars become the raw material for the next.

The leftover core, whether a neutron star or a black hole, remains as a kind of monument to the event. It carries the memory of its former star, but it behaves in a completely different way. Neutron stars, though stable, can still produce bursts of radiation or merge with others in titanic collisions that send gravitational waves rippling across the cosmos. Black holes, in turn, become hidden engines of the universe —feeding on nearby matter, shaping galaxies, and warping time itself. Both are legacies of the same explosive moment when a star's fuel finally ran out.

DIFFERENT SIZES OF BLACK HOLES

Not all black holes are created equal. Some are small enough that their entire event horizon could fit within a city. Others are so massive they could swallow our entire solar system without slowing down. Yet they're all made from the same basic process: gravity crushing matter into a region so dense that nothing can escape. What changes is how much matter you start with and how long the black hole has been around to gather more. Once scientists began to detect and measure them, they realized that black holes come in a surprising range of sizes—from stellar-sized remnants of dying stars to the supermassive giants that dominate the centers of galaxies.

The most common type found in the universe is the *stellar black hole*. These are born from the collapse of massive stars at the end of their lives. When a star several times larger than the Sun runs out of nuclear fuel, its core can no longer resist gravity, and the collapse becomes unstoppable. What's left behind after the explosion of a supernova is a compact remnant—a black hole a few times the mass of the Sun, but packed into a space barely a few miles across. To put that in perspective, if Earth were compressed into a black hole, it would be the size of a marble. A stellar black hole is the same idea, just on a much grander scale.

Stellar black holes are scattered throughout galaxies, often in binary systems where one normal star orbits the black hole. In these pairs, gas from the companion star can spiral toward the black hole, forming a bright, glowing disk as it heats up. Even though the black hole itself is invisible, that hot disk shines intensely in X-rays and light, revealing the hidden object. These systems allow astronomers to calculate the black hole's mass and confirm that what they're observing isn't just a dense star or cluster but a true region of no return.

But stellar black holes are only the beginning of the story. As scientists continued studying the cosmos, they found evidence of black holes that defied these smaller scales. At the centers of nearly all large galaxies—including our own Milky Way—lie *supermassive black holes*, millions or even billions of times more massive than the Sun. These colossal objects sit quietly most of the time, but when they feed on surrounding gas and dust, they can become the brightest things in the universe. Their gravitational fields are so strong that entire stars can orbit them safely without falling in.

The black hole at the center of our galaxy, called Sagittarius A*, is about four million times the mass of the Sun. Yet its event horizon—the boundary beyond which nothing escapes—would fit inside the orbit of Mercury. In other galaxies, the central black holes are even larger. Some weigh ten billion solar masses or more, with event horizons that could engulf our entire solar system and still have room to spare. Despite their incredible size, they obey the same laws as their smaller cousins. What changes is simply the scale: the same physics, stretched to cosmic proportions.

How these supermassive black holes form is still one of astronomy's biggest puzzles. A single collapsing star can't make something that large—it would take far too much mass. One idea is that they began as smaller black holes in the early universe that grew over billions of years, feeding on surrounding gas or merging with others. Another theory suggests that they formed directly from enormous clouds of gas collapsing before stars even existed. However they began, once they reached a certain size, they became anchors for galaxies. Their gravity helps hold billions of stars in orbit, shaping the structure of the galaxies we see today.

Between these two extremes—stellar and supermassive—there appears to be a middle ground, though it's harder to find. These are known as *intermediate-mass black holes.* They might range from hundreds to hundreds of thousands of times the mass of the Sun. Detecting them has proven tricky because they're too big to form in a single supernova and too small to sit at the centers of galaxies, where we'd easily notice them. Scientists have found hints of them in star clusters and in places where unusually bright bursts of X-rays suggest a black hole that's neither small nor colossal. If confirmed, they could help explain how supermassive black holes grew so quickly in the early universe—perhaps by merging many intermediate-sized ones together.

There's also a more speculative category: *primordial black holes.* These would be relics from the earliest moments after the Big Bang, formed not from stars but from the intense density fluctuations in the newborn universe itself. They could be as small as an atom or as large as a mountain, depending on the conditions when they formed. None have been detected yet, but if they exist, they might make up some of the mysterious "dark matter" that fills the universe. Their discovery would not only expand the size range of black holes but also rewrite what we know about the universe's first seconds.

The scale difference between these types of black holes is staggering. A stellar black hole could fit comfortably inside Manhattan, while a supermassive one might span the orbit of Neptune. Yet despite that difference, both curve space in exactly the same way—they just do it over different distances. If you could safely hover near a stellar black hole, the pull of gravity would feel intense and localized. Near a supermassive black hole, the pull at the event horizon would actually feel gentler, even though the total gravity is far greater. That's because the curvature of space is spread over a much larger area. It's a strange but true fact: the bigger the black hole, the less extreme the tidal forces near its edge.

That variation in size also affects how black holes evolve over time. Smaller ones, especially those formed from stars, can slowly lose mass through a process called Hawking radiation, a faint trickle of energy that leaks from their boundaries due to quantum effects. Over unimaginable timescales —longer than the current age of the universe—they could eventually evaporate entirely. Supermassive black holes, on the other hand, lose energy so slowly that they'll survive long after every star has burned out. In the far future, when galaxies have gone dark, these giants will be among the last active objects in the cosmos.

Understanding the range of black hole sizes helps scientists piece together the broader story of the universe's structure. Each type plays a different role. Stellar black holes are the endpoints of massive stars. Intermediate ones might bridge the gap, explaining how the first galactic cores formed. And supermassive black holes define the architecture of galaxies, controlling how stars and gas move on vast scales. Even though they all share the same defining feature —gravity so strong that light cannot escape—their effects are felt across very different corners of the cosmos.

3

DIFFERENT TYPES OF BLACK HOLES

STELLAR BLACK HOLES

Every black hole begins with gravity, but stellar black holes tell the most personal version of that story. They are the aftermath of stars that once burned brilliantly, lived fiercely, and ended in collapse. When a massive star reaches the end of its life, it leaves behind a reminder of just how powerful and unforgiving nature can be. These black holes are among the most common in the universe, yet they are anything but ordinary. They're the small but mighty fossils of cosmic giants.

To understand what makes a stellar black hole, it helps to start with the kind of star capable of becoming one. Most stars in the universe, including our Sun, will never reach that fate. The Sun, though bright and long-lived, simply isn't heavy enough. After billions of years of shining, it will end as a white dwarf—a dense, cooling ember of what it once was. But a star at least three times heavier than the Sun lives a different story. Its mass gives it tremendous power and, eventually, an equally tremendous downfall.

Throughout its life, a massive star burns through its nuclear fuel much faster than a smaller one. The immense gravitational pressure inside such a star pushes its core to higher temperatures and densities, speeding up the fusion process. It shines far more brightly, but at the cost of burning out in just a few million years—a blink of an eye in cosmic time. When it finally runs out of fuel, its core can no longer generate enough energy to resist the pull of gravity. Everything collapses inward, and in that moment, the star's death gives birth to something new: a black hole.

The collapse is so sudden and extreme that it defies intuition. The material that once formed the heart of a star larger than the Sun is crushed into a space no bigger than a city. The outer layers are blasted away in a supernova explosion, while the remaining core is compacted beyond the limits of ordinary matter. This tiny remnant—the stellar black hole—contains all the star's mass but none of its light. It becomes a gravitational trap from which nothing, not even the photons that once made the star shine, can escape.

The newly formed black hole doesn't sit idle. Its intense gravity begins shaping its surroundings almost immediately. If there's still gas and debris left over from the explosion, it can fall inward, heating up as it spirals toward the event horizon. The process releases enormous energy, creating a brilliant X-ray glow. Astronomers use that glow as a clue, because the black hole itself remains invisible. In binary systems—where a black hole orbits a companion star—this effect is even more dramatic. The black hole can draw matter from its partner, forming a swirling disk that lights up space with bursts of radiation.

These binary systems are among the best ways to find stellar black holes. When gas from the companion star falls in, it doesn't do so silently. It emits powerful X-rays and sometimes produces jets—narrow streams of particles moving at nearly the speed of light. These outbursts tell astronomers that something compact and extremely dense must be lurking there. By measuring how the companion star moves, scientists can calculate the mass of the invisible object. If it's too heavy to be a neutron star, the only explanation left is a black hole.

One of the fascinating aspects of stellar black holes is how varied they can be. Their masses typically range from a few times that of the Sun to a few dozen, depending on the original star. But recently, astronomers have found examples that challenge those limits—black holes with masses of fifty or more Suns, likely formed from stars that lost little mass before collapsing. These discoveries are reshaping our understanding of how black holes form and how different conditions, such as metallicity (the elements a star contains), can influence their final size.

Not all stellar black holes form in isolation. When two massive stars live together as a pair, they can both eventually collapse into black holes. Over time, their mutual gravity pulls them closer, and when they finally collide, the merger releases ripples in the fabric of space-time—gravitational waves. These faint waves were first detected in 2015, confirming a prediction made by Einstein a century earlier. Each detection since then has revealed new details about stellar black holes: their masses, how fast they spin, and even clues about where they formed.

The existence of stellar black holes also raises intriguing questions about how many there might be. Our galaxy alone could contain millions, scattered invisibly among the stars. Most are quiet, giving away their presence only through their gravitational pull. Some may drift through space alone, the remnants of stars that once had companions but were ejected by supernova blasts. Others remain tethered to binary systems, siphoning material from nearby stars and revealing themselves in X-ray outbursts.

Even though they're among the smallest types of black holes, stellar ones can be surprisingly active. When they feed on gas or dust, they convert that material into energy with remarkable efficiency. A black hole can turn as much as 10 percent of the falling matter's mass into energy—a level far beyond what nuclear fusion in stars can achieve. That's why the areas around black holes often shine brighter than the galaxies that contain them. Yet the black hole itself remains perfectly dark. The light comes not from within, but from matter's final moments before crossing the event horizon.

Studying stellar black holes also helps scientists understand the conditions that lead to more massive ones. Over billions of years, small black holes can merge to create larger ones, or they can sink toward galactic centers, feeding the growth of supermassive black holes. In that way, these stellar remnants might be the seeds from which cosmic giants are born. Every black hole that forms from a dying star adds a piece to that larger puzzle, showing how gravity builds complexity from simple beginnings.

SUPERMASSIVE BLACK HOLES

If stellar black holes are the embers left behind by dying stars, supermassive black holes are something entirely different—ancient engines of gravity that define the structure of galaxies themselves. These giants, weighing millions or even billions of times as much as the Sun, sit quietly at the heart of nearly every large galaxy known. They are not rare, and they are not exotic outliers. They are central players in the architecture of the universe. Their presence is so consistent that scientists now see them as part of what makes a galaxy *a galaxy* in the first place.

The black hole at the center of our own Milky Way is one of them. It's called Sagittarius A*, and despite its incredible mass—about four million Suns—it occupies a region smaller than the orbit of Mercury. It's quiet most of the time, hidden behind clouds of dust and gas, but occasionally it flares, releasing bursts of energy when a stray clump of matter drifts too close. Other galaxies harbor even more massive versions. In some distant galaxies, these giants weigh ten billion solar masses or more, their event horizons stretching across distances larger than our entire solar system.

The mystery that surrounds supermassive black holes begins with how they formed. Unlike stellar black holes, which are born from collapsing stars, these cosmic giants likely came into being through different processes. One theory suggests they started as smaller black holes that gradually grew by feeding on gas, dust, and stars over billions of years. Another idea is that they formed early in the universe's history, when massive clouds of primordial gas collapsed directly into black holes without forming stars first. However it happened, their growth seems to have gone hand in hand with the formation of galaxies themselves.

In fact, the relationship between galaxies and their central black holes is one of the most intriguing findings in modern astronomy. The size of a supermassive black hole is closely linked to the size of the galaxy's central bulge—the dense cluster of stars near the middle. This connection suggests that galaxies and their black holes evolve together. As the galaxy grows, so does its black hole, and vice versa. The energy released by the black hole as it feeds can influence how new stars form, regulating the galaxy's development like a kind of cosmic feedback system.

When a supermassive black hole feeds actively, the results can be astonishing. Gas and dust falling toward it form a vast, hot accretion disk. As this material spirals inward, friction and magnetic forces heat it to millions of degrees. That energy escapes as intense radiation, often outshining the entire galaxy around it. These bright, active black holes are called *quasars*. From billions of light-years away, quasars appear as tiny points of light, yet they can emit more energy than all the stars in a galaxy combined. Some have been observed from when the universe was less than a billion years old, meaning they formed astonishingly fast—another clue that their origins may differ from the smaller black holes we know.

In addition to the brilliant light of quasars, supermassive black holes can unleash enormous jets of particles and energy. These jets can stretch for thousands of light-years, punching through the galaxy's surrounding gas and even affecting neighboring galaxies. The jets are powered not by the black hole itself, but by magnetic fields in the swirling plasma around it. As the material near the event horizon twists and heats, some of it gets flung outward along the poles at nearly the speed of light. These jets can carry more energy than entire clusters of galaxies emit in starlight, reshaping the regions they pass through.

Yet, despite their violent potential, supermassive black holes also have calm, long-lasting phases. Once a black hole has consumed most of the nearby material, it quiets down, entering a dormant state. That's likely what's happening in our own galaxy today. Sagittarius A* emits only faint radiation, and the stars near it orbit in stable patterns, their motions revealing the hidden mass at the center. The quiet doesn't mean it's inactive; it just means it has nothing left to consume nearby. If a passing gas cloud or unlucky star wandered too close, it would flare to life again.

The influence of a supermassive black hole extends far beyond its immediate surroundings. Though the black hole itself is confined to the galactic center, its gravity helps organize the motions of billions of stars. In some cases, entire galaxies rotate in harmony around their central mass. But gravity isn't the only way it shapes its environment. The radiation and winds from an active black hole can heat surrounding gas, preventing it from cooling and condensing into new stars. This process can control the rate of star formation in the galaxy, keeping it from producing stars too quickly. In this way, black holes help set the pace of galactic evolution.

The scale of these objects defies easy comprehension. A black hole with a billion times the Sun's mass would have an event horizon roughly the size of our solar system, yet it could sit at the core of a galaxy tens of thousands of light-years across. From that small region, it can influence structures spanning hundreds of thousands of light-years. The contrast between the black hole's compactness and its reach is one of the great paradoxes of the cosmos: the smallest volumes can contain the greatest gravitational power.

Despite their size and influence, supermassive black holes are surprisingly simple in their measurable properties. Like their smaller relatives, they can be described by just three main characteristics: mass, spin, and electric charge. Of these, mass and spin are the most significant. The spin, or angular momentum, tells scientists how fast the black hole is rotating. Some spin at nearly the maximum rate allowed by physics, dragging the space around them in a phenomenon called "frame dragging." This rotation can shape how the accretion disk forms and how jets are launched.

Observing supermassive black holes directly is an enormous challenge. They don't emit light, and the gas around them often obscures the view. But in recent years, new technologies have given scientists their first glimpses. In 2019, the Event Horizon Telescope collaboration released the first image of a black hole's shadow—a dark circle surrounded by glowing light—from the galaxy M87. The object is six and a half billion times the mass of the Sun, and the image showed what was once only theoretical: the bright ring of light from matter swirling just outside the event horizon. That observation confirmed decades of predictions and opened a new window into studying black holes directly.

Supermassive black holes also merge. When two galaxies collide, their central black holes eventually find each other and spiral inward, locked in a gravitational dance. As they orbit and merge, they emit gravitational waves—ripples in space-time that carry energy across the universe. These events are slower and more powerful than the mergers of smaller black holes, producing waves with much longer wavelengths. Detecting them requires sensitive instruments and long-term observations, but their discovery would provide another key to understanding how galaxies grow and evolve over cosmic time.

INTERMEDIATE BLACK HOLES

Between the tiny but fierce stellar black holes and the monstrous supermassive ones that anchor galaxies, there's a puzzling gap. Astronomers have long expected that something should exist in the middle—a class of black holes not small enough to be made from dying stars, but not vast enough to sit in the centers of galaxies. These would be the *intermediate black holes*, sometimes called the "missing link" of cosmic evolution. For decades, they existed mostly as theory, an idea waiting for proof. Yet hints keep emerging that they may indeed be out there, quietly bridging two extremes.

The name "intermediate" might sound unremarkable, but it represents a profound question about how black holes grow. If the universe has both small black holes, formed from individual stars, and gigantic ones that rule galaxies, how did it make the jump from one to the other? It's unlikely that a few stellar black holes simply ballooned into supermassive ones overnight. Growth in the cosmos takes time and process, and intermediate black holes may be the missing steps in that ladder—ancestors of the cosmic giants.

These mid-sized black holes are thought to weigh anywhere from a few hundred to a few hundred thousand times the mass of the Sun. They're heavy enough to outstrip anything created by a single star's collapse, but not nearly as enormous as the supermassive titans found in galactic centers. The question has always been how such objects could form in the first place. If they're too big to come from one star, what could have produced them?

One idea is that they're built gradually, through the mergers of smaller black holes and dense star clusters. In crowded regions of space—such as globular clusters, where hundreds of thousands of stars orbit in close quarters—stellar black holes might collide and merge over time. Each merger creates a heavier black hole, which in turn attracts more companions, continuing the cycle. Given enough time, this process could produce an object that's hundreds or thousands of solar masses—a natural stepping stone toward the colossal black holes we see in galaxies.

Another possible route is through the direct collapse of massive gas clouds in the early universe. In the universe's first few hundred million years, conditions were much denser and more uniform than they are now. Large clouds of hydrogen and helium might have collapsed straight into black holes without forming stars first. These would start off far more massive than stellar black holes, making them perfect candidates to grow into intermediate or even supermassive ones later. Evidence for this idea comes from how early we see bright quasars—powered by enormous black holes—in the young universe. Something must have helped them grow fast, and direct-collapse black holes could be that head start.

For all these theories, finding one of these middleweight black holes has proven extremely difficult. They're rare, and most of the time they're quiet. A black hole without anything to feed on is invisible. It doesn't glow, radiate, or emit X-rays. Unless it happens to disturb its surroundings in some way, it's effectively undetectable. The only way to spot one is through its gravitational influence—how it bends light, tugs on stars, or disturbs the motion of gas.

The first strong hints came from studying star clusters. Astronomers found that in some of these dense, spherical gatherings of ancient stars, the orbits of certain stars near the center moved as though something very heavy and compact was lurking there. The object wasn't massive enough to be a supermassive black hole, but it was far too heavy to be a collection of ordinary stars. These findings pointed to the possibility of intermediate-mass black holes hidden in the hearts of star clusters, quietly exerting their pull.

Another clue came from unusual bursts of radiation observed in space. Some galaxies produce flashes of X-rays that are brighter than what a single stellar black hole could generate but dimmer than the blinding output of a supermassive one. These events are called *ultraluminous X-ray sources*. Many scientists suspect they're powered by intermediate black holes consuming material—perhaps a passing star that strayed too close or a stream of interstellar gas drawn in by gravity. The brightness of these sources suggests an accretion process more powerful than typical stellar systems, giving more weight to the idea that midsized black holes really exist.

In recent years, gravitational wave astronomy has added new evidence. When two black holes merge, the ripples they send through space-time carry information about their masses. Some of the detected signals hint at mergers that produced objects larger than typical stellar black holes but still smaller than supermassive ones. These events could represent the birth of new intermediate black holes—freshly formed from the collision of smaller ones. While not direct proof of long-lived intermediate black holes, these observations make the case stronger that the universe can and does create them.

If intermediate black holes do exist in significant numbers, they may hold the key to understanding how galaxies evolve. It's possible that many galaxies began with one or several of these objects, which then merged and grew over billions of years into the supermassive black holes we see today. They could also be the hidden occupants of small or faint galaxies that never developed enough mass to grow giants. Dwarf galaxies, for example, may contain intermediate black holes quietly shaping their structure from within.

Their existence also carries implications for how the early universe behaved. If intermediate black holes formed from dense gas clouds soon after the Big Bang, they could help explain how the first galaxies formed so quickly. These black holes would act as gravitational seeds, pulling matter toward them and helping to organize it into galaxies. The feedback from their energy output—jets, winds, and radiation—would then influence how stars formed around them, setting patterns that continue today.

Because they're so difficult to detect, intermediate black holes remain one of the most intriguing mysteries in astrophysics. The search for them continues through a mix of indirect methods. Astronomers look for subtle distortions in starlight caused by gravitational lensing, where a massive object bends the light from something behind it. They monitor dense star clusters and dwarf galaxies for signs of invisible mass at their centers. They track the faint whispers of gravitational waves for patterns that reveal heavier-than-expected mergers. Each new observation adds a piece to a picture that's slowly coming into focus.

The reason this hunt matters goes beyond curiosity. Intermediate black holes could provide a missing chapter in the story of how structure forms in the universe. Without them, it's hard to explain how supermassive black holes reached such enormous sizes so early in cosmic history. It's also hard to connect the tiny black holes formed by stars with the behemoths that shape galaxies. These middleweights offer a bridge—a way to make sense of how gravity scales from the collapse of a single star to the architecture of an entire galaxy.

They might also help explain some of the universe's most violent events. If intermediate black holes are hiding in star clusters, they could occasionally collide with other black holes, sending out bursts of gravitational waves unlike anything yet observed. They might also play a role in shaping the dynamics of galaxies by capturing stars and redistributing energy within their environments. Every piece of evidence points to the idea that these objects, while elusive, are important threads in the fabric of cosmic evolution.

PRIMORDIAL BLACK HOLES

Before there were stars, before galaxies took shape, before even light had room to move freely, the universe was a dense, boiling sea of energy. Out of that chaos, some scientists think black holes might have been among the very first structures to form. These hypothetical objects, called *primordial black holes*, would have appeared in the first tiny fractions of a second after the Big Bang. They are unlike the black holes formed later from collapsing stars. Instead of being born from death, they would have been born from the very beginning—fossils of creation itself.

The early universe was nothing like what we see today. It was incredibly hot, dense, and uniform, yet not perfectly smooth. Tiny fluctuations in density—regions that were just slightly more crowded with matter than others—served as the seeds from which all later structures grew. Over billions of years, those slight variations became galaxies, stars, and planets. But in that first instant, when conditions were extreme beyond comprehension, even a small excess of density could have collapsed under its own gravity, forming a black hole directly. That's the idea behind primordial black holes: they didn't need stars to die first. The universe itself provided the conditions for their birth.

The scale of such objects could vary widely. Depending on when and where they formed, primordial black holes could range from microscopic specks to the mass of mountains—or even as heavy as the Sun. In the earliest fractions of a second, the density of the universe was enormous, meaning that even a tiny region could contain enough mass to collapse. As the universe expanded and thinned out, the opportunity for such collapses diminished. That means primordial black holes, if they exist, could tell us a lot about what happened in that short, mysterious window after the Big Bang.

What makes them fascinating is how they connect the smallest scales of physics to the largest. In that early moment, quantum effects—the rules that govern the subatomic world—interacted directly with gravity, the force that shapes galaxies. Normally, these two realms don't overlap in everyday conditions, but the early universe was a special case. It was a laboratory where matter, energy, and gravity were entangled in ways we can barely describe. Primordial black holes, if they were created then, might preserve evidence of those conditions, frozen in time.

Unlike stellar or supermassive black holes, primordial ones wouldn't have been limited by size or process. They could have formed in enormous numbers, scattered everywhere throughout the cosmos. Some might have evaporated long ago through a process known as Hawking radiation—a slow leak of energy that allows the smallest black holes to shrink over time. Others, if massive enough, could still exist today, drifting silently through space. The idea that countless tiny black holes might be scattered across the universe is both thrilling and unsettling. They would be nearly impossible to detect, yet collectively, they might play a major role in shaping cosmic history.

One of the most intriguing possibilities is that primordial black holes could account for at least part of the mysterious *dark matter* that fills the universe. Dark matter makes up most of the mass in galaxies, yet it neither emits nor reflects light, and no one knows what it is. If primordial black holes are real and numerous, they could be that missing component—massive, invisible, and scattered everywhere. This idea has inspired decades of research, with scientists searching for signs that could confirm or disprove it.

Detecting something that doesn't emit light is challenging, but nature sometimes helps out. One way scientists search for primordial black holes is through *gravitational lensing*. Gravity bends light, and a compact object like a black hole can act as a lens, distorting or magnifying the light from a star or galaxy behind it. If a primordial black hole passes in front of a distant star, the star's brightness might flicker or intensify for a short time. Observing these subtle effects could reveal whether such small black holes exist.

Another line of investigation comes from the study of cosmic background radiation—the faint afterglow of the Big Bang that fills the universe. Primordial black holes would have influenced how matter and radiation interacted in the early universe, leaving subtle imprints on that background. By studying those patterns in extraordinary detail, scientists can test whether black holes of certain sizes could have formed without disrupting what we see today. So far, the results haven't ruled them out completely, but they've narrowed the possibilities.

Gravitational wave observatories have added another tool to the search. When two black holes merge, they release ripples in space-time that can travel across the cosmos. Some of the black holes detected through these waves are unusually heavy for their kind—too large to have formed easily from the collapse of stars. One hypothesis is that these might be primordial black holes, ancient objects that merged after billions of years of wandering. If that's true, gravitational wave detections could be our first indirect glimpse of them.

The concept of primordial black holes also opens up questions about the ultimate limits of physics. If they really did form at the dawn of time, they might contain clues about how gravity behaves at the smallest scales—something current theories can't fully explain. Our best model for gravity, general relativity, describes how massive objects curve space-time, but it doesn't blend easily with quantum mechanics, which governs the behavior of particles. The conditions that would create primordial black holes lie right at that intersection. Understanding them could help unite these two pillars of physics into a single framework—a "theory of everything" that scientists have sought for decades.

Some researchers have even proposed that the universe we see might have been influenced by primordial black holes in unexpected ways. For example, their presence could have triggered the formation of the first stars or galaxies by acting as gravitational seeds that gathered gas and dust around them. Others suggest they might explain certain high-energy cosmic events we observe today, such as bursts of gamma rays or unusual patterns in radiation. Whether or not these ideas prove correct, they show how rich the concept of primordial black holes has become—a crossroads of cosmology, particle physics, and relativity.

If primordial black holes exist, they would be the oldest surviving structures in the universe, predating stars, planets, and galaxies. They would have witnessed everything: the first atoms forming, the first light shining, the clustering of matter into galaxies, and the slow expansion of the cosmos into what we see today. Their mere presence would mean that gravity had been shaping the universe from the very first moment, sculpting order out of chaos.

Yet, for all their theoretical appeal, the evidence remains elusive. Every test so far has come up just shy of confirmation. The difficulty lies in their nature—they don't shine, they don't radiate (unless very small), and they may be spread thinly across vast distances. Detecting one would require extraordinary precision and luck. Still, science often progresses through the search for such unseen things. Even if primordial black holes turn out not to exist, the effort to find them deepens our understanding of the universe's first seconds, forcing us to refine our models of how everything began.

UNDERSTANDING THEIR SCALES

The trouble with black holes isn't just that they're invisible —it's that their scales are hard to wrap your head around. Numbers like "millions of times the mass of the Sun" sound impressive, but they don't mean much to most people beyond "really big." To make sense of black holes, it helps to find comparisons that connect their immense differences in size and mass to something familiar. Think of it as trying to fit the vastness of the cosmos into a scale we can actually picture, one that brings a little order to something almost beyond comprehension.

Start small. A stellar black hole—the kind left behind by a collapsing star—is the compact remnant of something that once stretched millions of miles across. After the star explodes, all that material gets crushed into a sphere perhaps twenty miles wide. If Earth were reduced to a black hole, it would be about the size of a marble. That's the kind of density we're talking about—an entire planet's worth of matter packed into something you could hold in your hand. A stellar black hole is the same principle, just on a grander level: the mass of several Suns squeezed into a space no larger than a city.

Now stretch upward to intermediate black holes. These middleweights occupy a strange place between the ordinary and the colossal. If a stellar black hole is like a pebble on a beach, an intermediate black hole is a mountain. It has more weight, more pull, and a broader reach, but it's still dwarfed by what lies beyond. The event horizon of an intermediate black hole—its boundary of no return—might stretch hundreds or even thousands of miles across. It's vast compared to its smaller cousins, but in the cosmic scale, it's still a small island in an enormous sea.

To get a sense of what that means, imagine scaling up by orders of magnitude. If a stellar black hole could fit within the footprint of a large city, then an intermediate one would fill a continent. And yet both would be invisible, their gravity defining their presence rather than any visible edge. It's one of those strange realities of physics: size doesn't directly determine presence. A small black hole can be more violent in its pull near the event horizon than a larger one, even though the larger exerts greater gravity overall.

Then there are the supermassive black holes—the planetary, even galactic, scale versions of this same phenomenon. A supermassive black hole can contain the mass of millions or billions of Suns, yet it's still contained within a relatively compact region of space. If we were to extend the analogy, a stellar black hole might be a pebble, an intermediate black hole a mountain, and a supermassive black hole an entire planet—or, in the most extreme cases, a world so large it could swallow the planet whole and still have room to spare.

For perspective, the supermassive black hole at the center of the Milky Way, Sagittarius A*, is about four million times the mass of the Sun, with an event horizon roughly the size of Mercury's orbit. The one in the galaxy M87 is more than a thousand times larger—big enough that light itself takes nearly two days to travel across its diameter. If you tried to compare them to something familiar, you'd have to stretch the analogy from small pebbles to massive worlds. A stellar black hole is a hand-held stone; an intermediate black hole, a mountain chain; a supermassive one, a planet or even a star system in scale. Each step up multiplies mass and size in ways that break human intuition.

These differences also change the way black holes behave. Near a small stellar black hole, gravity changes abruptly. The pull increases so sharply that anything approaching it is stretched into long, thin strands in a process called spaghettification. Near a supermassive black hole, though, that gradient is gentler. The gravitational pull is still enormous, but it's spread over a much larger area. You could, in theory, cross the event horizon of a supermassive black hole without being torn apart immediately—though you'd still never escape once inside. The point is that even though these objects differ enormously in scale, the physics that govern them are exactly the same. The only thing that changes is the degree.

What's remarkable about black holes is how their scale defies normal comparisons. In everyday life, when something gets larger, it usually grows heavier in a way we can predict. A boulder weighs more than a pebble; a mountain weighs more than a boulder. But black holes compress that logic. A black hole ten times the mass of another isn't just ten times heavier—it's often far more extreme in density and gravitational power. And when you step up from a stellar black hole to a supermassive one, the difference isn't gradual; it's like leaping from a pebble to an entire continent.

To help visualize this, think in terms of escape velocity —the speed something must reach to break free from gravity. On Earth, that speed is about 25,000 miles per hour. On a neutron star, it's close to half the speed of light. On the edge of a stellar black hole's event horizon, you'd need the full speed of light to escape—and still fail. Now take that same rule and apply it to larger and larger black holes. The event horizon grows, but the condition remains the same: once you cross it, escape becomes impossible. No matter how large or small, every black hole's defining boundary is where light itself becomes trapped.

These scales also shape how long black holes live. The smaller ones, particularly the tiniest theoretical ones—perhaps even microscopic primordial black holes—could evaporate over time through Hawking radiation. Larger ones, like stellar and intermediate black holes, would last far longer, their lifespans measured in numbers that stretch beyond the age of the universe. The giants—supermassive black holes—could persist almost indefinitely, outlasting stars, galaxies, and perhaps even the expansion of the universe itself. It's a strange hierarchy: the larger the black hole, the more stable it becomes. Like comparing a droplet of water to an ocean, the bigger it gets, the harder it is to disturb.

These comparisons highlight something profound about scale in the universe: growth isn't just about size—it's about influence. A pebble sits unnoticed. A mountain shapes weather. A planet holds moons in orbit. A galaxy bends the paths of light itself. In that same way, a small black hole influences its immediate surroundings, perhaps a few nearby stars or a companion. An intermediate one can affect clusters of stars. But a supermassive black hole holds sway over entire galaxies, its gravity helping determine how billions of stars move and how new ones form.

Even their formation histories can be thought of in terms of scale. A stellar black hole represents a local event—the death of a single massive star. An intermediate one represents cooperation, where many such remnants merge and accumulate over time. A supermassive black hole represents cosmic architecture—structures so large and interconnected that they define the framework of galaxies themselves. Each scale adds a new level of complexity, from solitary collapse to collective growth to structural dominance.

Yet no matter the scale, all black holes share an extraordinary balance between simplicity and mystery. They are described by only a few numbers—mass, spin, and charge—but those numbers translate into vastly different realities. A black hole ten times the mass of the Sun is small, fierce, and violent. One a billion times that mass is stately and vast, capable of shaping galaxies without flaring brightly. From pebble to mountain to planet to galaxy, the essence stays the same: a region of space where gravity has gone too far.

4

WHAT'S INSIDE (AND WHAT ISN'T)

When people talk about black holes, they often speak as if the entire thing is one mysterious object—a single, featureless void. But a black hole isn't uniform. It has structure, though not in the sense of solid layers or visible features. It's more like a system made up of distinct zones, each defined by how gravity behaves there. If you break it down, three main components describe nearly everything that matters about a black hole: the *event horizon*, the *singularity*, and the *accretion disk*. Together, these parts explain how a black hole looks, how it interacts with its surroundings, and what happens to anything that ventures too close.

Start with the **event horizon**, because that's the boundary everyone has heard about—the famous "point of no return." It isn't a surface made of matter, nor does it glow or burn. It's a region in space where gravity's pull becomes so strong that the escape velocity equals the speed of light. Anything crossing that threshold—matter, energy, or even light itself—cannot get back out. If you think of the universe as a vast ocean, the event horizon is where the current becomes unstoppable, dragging everything inward.

From an outside perspective, nothing ever appears to cross the event horizon. As an object falls closer, its light takes longer and longer to reach us, stretched into redder wavelengths by the gravitational field. It slows, dims, and eventually fades away. The object seems to freeze at the edge, suspended forever on the brink. But from the object's own point of view, it crosses the event horizon in a finite time, continuing its fall into darkness. This difference between outside and inside perspectives highlights how drastically gravity can warp time.

Once inside, the rules of physics shift beyond recognition. The event horizon marks the limit of what can be observed, not just practically but fundamentally. Information cannot escape from beyond it, which means the rest of the universe can never know what happens inside. The horizon, in that sense, isn't merely a boundary in space—it's a boundary in knowledge. It divides what can be known from what can never be retrieved.

The horizon's size depends on the black hole's mass. For a stellar black hole, the event horizon might be only a few miles across. For a supermassive one, it could be wider than our solar system. Despite that difference, the concept is the same: it's the place where gravity takes full command, and the familiar laws of motion no longer apply. The larger the black hole, the gentler the curvature at its horizon. In a small one, the difference in gravitational pull between your head and your feet would tear you apart instantly. In a supermassive black hole, the same crossing might feel smooth and uneventful—at least initially.

Deeper inside lies the **singularity**, the center point where all paths converge. This is where the equations of general relativity predict that matter collapses into infinite density and zero volume. That idea, while mathematically neat, defies physical reality as we know it. Infinity isn't something nature usually allows; it's a sign that our theories have hit their limits. The singularity is the place where our understanding breaks down completely. We can't describe what happens there using any existing model of physics.

Some scientists believe that the singularity isn't a point at all but rather a region where quantum effects—those that dominate at the smallest scales—counteract gravity's collapse. It might be a compressed core of exotic matter or even a bridge to another part of space-time. These ideas remain speculative, but they point to one essential truth: whatever exists at the heart of a black hole, it's beyond what current science can fully describe. The singularity, in that sense, isn't just a physical concept—it's a frontier of human knowledge.

Despite its theoretical elusiveness, the singularity shapes everything around it. It's the source of the black hole's gravitational field, the reason the event horizon exists. All the matter and energy that fall in contribute to it, adding to the black hole's total mass. Yet from the outside, we can't tell anything about what's inside. Two black holes of the same mass, spin, and charge are indistinguishable, no matter what they were made from. That uniformity—sometimes called the "no-hair theorem"—is one of the most striking features of these objects. All individuality is lost; only the fundamental properties remain.

If the event horizon defines the boundary and the singularity defines the core, the **accretion disk** is the part that makes black holes visible. Despite the name, a black hole itself doesn't emit light. What we see comes from the material swirling around it before it's consumed. Gas, dust, and sometimes even stars spiral inward under gravity, forming a flat, rotating disk of matter. As this material falls closer to the event horizon, it speeds up and compresses, heating to millions of degrees. The result is an intense glow, often visible across enormous distances in X-rays and other high-energy light.

The accretion disk is where most of the action happens. It's chaotic, turbulent, and incredibly bright. Magnetic fields twist through it, channeling some of the energy outward in enormous jets that can shoot thousands of light-years into space. These jets aren't made of material escaping the black hole itself—they come from the heated plasma and magnetic forces near the outer edge of the disk. The black hole's spin can also influence them, shaping how the energy is released. In the centers of active galaxies, these effects can make black holes the brightest objects in the universe, even though the black holes themselves remain dark.

What's fascinating about the accretion disk is that it's both destructive and creative. It feeds the black hole, but in the process, it also generates light, heat, and radiation that influence everything nearby. In galaxies with active black holes, the energy from accretion can shape how stars form and how gas moves. It's a reminder that even the most extreme objects in the cosmos participate in cycles of transformation. The same gravity that devours matter also powers the processes that illuminate the universe.

For smaller black holes, the accretion disk can be fed by a companion star. Gas from the star is drawn off and forms a glowing spiral around the black hole. These systems, known as X-ray binaries, are among the best ways to detect black holes. The radiation from their accretion disks reveals how fast matter is moving, how hot it becomes, and how strong the gravitational field must be. Each observation is like watching the last moments of matter's existence before it disappears forever.

The interplay between the event horizon, the singularity, and the accretion disk captures the essence of what a black hole is. The disk shows us how black holes reveal themselves, the horizon marks the limit of what we can know, and the singularity represents the ultimate mystery beyond that limit. Each layer connects to the others, forming a structure that is both simple and incomprehensible—simple in its geometry, incomprehensible in its extremes.

One of the most interesting things about these components is how they illustrate the relationship between what can be seen and what can't. The accretion disk shines brilliantly, yet everything that makes it glow is doomed to vanish. The event horizon is invisible, yet it's the most defining feature of all. The singularity, the true center, is beyond all reach, hidden behind layers of physics that no instrument can penetrate. Together, they form a kind of cosmic paradox: an object defined entirely by what it hides.

THE "POINT OF NO RETURN"

The phrase *"point of no return"* perfectly captures what makes black holes so mysterious and fascinating. It refers to the event horizon—the invisible boundary surrounding every black hole where gravity becomes so powerful that escape is impossible. Nothing, not even light, can get back out once it crosses that line. While that concept might sound abstract, there are plenty of ways to picture it using experiences we already understand. The universe may deal in physics, but our daily lives are full of smaller versions of the same principles.

Think first about a **waterfall.** Picture a calm river

flowing toward the edge. Upstream, the water moves lazily. You could row a boat there and easily paddle against the current to stay in place. But as you drift closer to the falls, the water speeds up. The current grows stronger, tugging harder at your boat. There's still a chance to escape if you start paddling early enough, but the closer you get, the faster the water pulls you forward. Then you reach the point where the current moves faster than you could ever row. Past that line, no amount of strength or effort can stop the inevitable. You're going over. That invisible border where turning back is no longer possible—that's what scientists mean by the event horizon. It's the cosmic equivalent of the edge of the waterfall.

Now, extend that metaphor further. The waterfall itself isn't dangerous to the river; it's part of its flow. The real danger lies in how the flow changes as you approach it. Gravity works much the same way near a black hole. Far from it, the pull is weak, like the gentle current upstream. Closer in, it becomes irresistible. There's no visible sign marking the edge, no flashing warning light, but the laws of motion guarantee what happens next. Once you pass that critical threshold, all paths lead inward.

A **steep hill** offers another way to understand the same idea. On a mild slope, you can walk upward, pause, or even roll a ball and watch it stop naturally. But the steeper the hill, the harder it becomes to climb back once you've started down. Eventually, you reach an angle where gravity wins—where no amount of effort can reverse your motion. The event horizon is like that ultimate slope, one so steep that every direction points downward. No matter how fast you could run or how powerful your rocket might be, you'd still be sliding toward the center.

Scientists often use another everyday comparison:

quicksand. The name itself sounds like science fiction, but the behavior is real. In ordinary mud, you can push off and move slowly forward. In quicksand, the more you struggle, the more you sink. The surface tension and pressure combine to trap anything that disturbs it. That's what crossing the event horizon is like—not because the black hole is sticky, but because every attempt to resist its pull only deepens the commitment. Once you're caught in its gravity, motion itself becomes your undoing.

But there's an important distinction. A person trapped in quicksand can still call for help. A boat drifting toward a waterfall might signal for rescue. In the case of a black hole, even communication stops at the event horizon. Light—the fastest thing in existence—can't escape once it's beyond that limit. That means information itself is trapped. To the rest of the universe, whatever crosses the horizon simply disappears. From an external viewpoint, it fades and slows until it's frozen in time, never actually vanishing but never returning either. It's as if the waterfall's edge stretched time itself, turning the final fall into an endless moment.

Another metaphor scientists sometimes use is **a stretched rubber sheet**, representing space-time. Place a heavy ball in the center of the sheet, and it sinks, creating a dip. Smaller balls rolling nearby spiral inward, trapped by the curve. If you make that dip infinitely deep, you've got the shape of a black hole's gravity. Near the center, the slope becomes so steep that not even light, which always moves at the cosmic speed limit, can climb back out. The event horizon is the ring around that pit where the slope becomes absolute. Once you pass it, there's no escape path, no upward climb left in the geometry of space itself.

This "point of no return" doesn't mean that everything inside instantly hits destruction. In fact, for large black holes, an object could cross the horizon without even feeling anything special at first. There's no solid surface or sudden jolt. The crossing would be silent and smooth, though it would also be final. Space-time's structure itself ensures that every direction now leads deeper inward, much like a river that bends and loops but always ends in the same drop.

One way to make this concept more relatable is by thinking about **time as part of the landscape**. Far from a black hole, time flows evenly, just as it does here on Earth. Near the event horizon, time behaves differently. To an outside observer, it slows down dramatically. The closer something gets to the horizon, the slower its clock appears to tick. At the horizon itself, it would seem to stop completely. From the falling object's perspective, though, time moves normally—it's space that becomes distorted. That reversal of roles between space and time is one of the strangest consequences of extreme gravity. You could think of it as trying to climb a hill where time itself becomes another direction—one that's always pushing you downward.

If we go back to the **waterfall** image, it's as though you're not just being pulled forward in space but also pulled forward in time at a rate you can't control. Every second brings you closer to the drop, and once you cross it, the future holds only one outcome. The event horizon doesn't destroy you directly—it defines what your future must be. It's not a wall; it's destiny made by gravity.

Even the idea of *light* struggling to escape can be understood in familiar terms. Think of a rocket launched upward from Earth. If it goes fast enough—about 11 kilometers per second—it breaks free from Earth's gravity. That's the escape velocity. For the Moon, it's smaller. For the Sun, it's much higher. For a black hole, that number surpasses the speed of light. Nothing can go faster than light, which means nothing can escape. The event horizon is where that mathematical line is crossed, where even light's best effort isn't enough. It's the ultimate limit built into the fabric of the universe.

The term "point of no return" makes it sound like there's a single spot where everything changes, but in reality, the pull grows gradually stronger the closer you get. The horizon simply marks the threshold where the curve of space becomes too steep. That idea helps explain why black holes aren't cosmic vacuum cleaners. They don't suck in everything nearby indiscriminately. Things can orbit a black hole safely at a distance, just like planets orbit stars. It's only when something drifts too close, past that invisible edge, that it becomes part of the one-way flow inward.

Another metaphor that works well is **a whirlpool in the ocean**. Water near the edge can circle safely, swirling but staying afloat. As you get closer, the current tightens, spinning faster and faster. There's a line beyond which the rotation is too strong to resist. The event horizon is like that line, the place where the spin of space-time drags everything down into the center. Even light, traveling straight, is bent into a spiral it can't escape. The image of water circling faster and faster before vanishing captures both the beauty and inevitability of what happens near a black hole.

WHAT WE *DON'T* KNOW

Every black hole hides two great questions that stretch the limits of science: what happens to the matter that falls inside, and what happens to time itself once it crosses the event horizon? These aren't the kinds of questions we can answer by pointing a telescope at the sky. They're questions that push against the very edge of physics—where our best theories start to blur.

When matter falls into a black hole, it doesn't simply vanish, but what becomes of it after it crosses the horizon is uncertain. To understand why, it helps to remember that everything we know about black holes comes from how they affect their surroundings, not from anything inside. The event horizon hides the interior from observation. Beyond that invisible shell, light and information can't escape, which means the rest of the universe is blind to whatever happens there. That makes black holes the ultimate test of our scientific imagination—real objects that force us to think about what can't be seen.

Let's start with the first part of the mystery: **where does the matter go?** When a star collapses into a black hole, all the atoms that once made up its core are crushed under their own gravity. The result is an object with a finite mass but a seemingly infinite density—a condition that makes no sense according to the laws of physics we currently know. Infinite density means the equations are giving us nonsense answers, a sign that they're being pushed past their breaking point. That's what scientists call the *singularity problem.*

According to general relativity, all the matter that falls into a black hole gets squeezed into that infinitely small region at the center. But if that's true, the question becomes: *what does "squeezed" even mean when space and time themselves lose their normal behavior?* In everyday physics, we can describe pressure, temperature, and movement because those ideas rely on time and distance. At the singularity, both collapse into something undefined. We can't describe where or when anything is. The math stops working there, which means our current theories simply run out of map.

This is why physicists often say black holes are where

general relativity and quantum mechanics come into direct conflict. Einstein's theory of gravity describes how massive things shape space-time. Quantum mechanics describes how matter behaves on tiny scales. Both work beautifully on their own—but not together. In the extreme environment of a black hole, both are needed, yet they give incompatible answers. That's why understanding what happens to matter inside a black hole requires a new kind of theory—something that can blend gravity with quantum physics.

Until that theory exists, scientists can only make educated guesses. Some believe that matter doesn't disappear but instead becomes part of the black hole's internal structure, its mass added to the whole. Others think that at the singularity, matter might be converted into something we don't yet understand—perhaps a kind of pure energy or a new state of matter entirely. There's even a radical idea that a black hole could act like a tunnel through space-time—a *wormhole*—where matter isn't destroyed but re-emerges somewhere else. It's a fascinating concept, though still speculative.

A more cautious possibility is that the matter simply ceases to exist in any meaningful sense within our universe. The information that once described it—its position, energy, composition—might be lost forever. But that idea creates another major problem known as the *information paradox*. According to the principles of quantum physics, information can never be destroyed. It can change form, spread out, or become impossible to read, but it can't vanish. Yet if black holes truly erase information, that rule would be broken. That contradiction has puzzled physicists for decades.

In recent years, some researchers have suggested that information might not be lost but somehow encoded on the black hole's surface—on the event horizon itself. In this view, everything that falls in leaves a kind of imprint, like data stored in a cosmic memory. The black hole's surface might then act like a vast information screen, preserving the details of everything it consumes, even if we can never recover them. It's a bold idea that ties into a wider concept called the *holographic principle*, which suggests that all the information about a region of space might be stored on its boundary. If true, it would mean that the universe itself behaves a bit like a hologram—three-dimensional reality emerging from two-dimensional information.

Now, to the second great question: **does time stop at the event horizon?** The answer depends on where you're standing. From far away, it looks as though time really does stop. As an object approaches the event horizon, its motion appears to slow. The light it gives off stretches into redder wavelengths, fading into darkness. To an outside observer, it never quite crosses the horizon; it just grows dimmer and fainter, frozen on the edge of eternity.

But from the point of view of the falling object, time doesn't stop at all. Everything seems normal. The crossing happens in a finite amount of time, and the fall continues inward. The difference comes from how gravity affects the flow of time. Near a black hole, gravity is so strong that it warps time itself, stretching it differently for different observers. What feels like seconds to one person could look like millions of years to someone far away. This isn't just theory—it's been measured around less extreme gravitational fields, like near planets and stars. Black holes are simply the most dramatic example.

At the event horizon, time behaves in such a way that the outside universe appears to speed up infinitely. To the falling object, the rest of the cosmos would seem to race through eons in an instant. Stars would age and die, galaxies would drift apart, and the universe itself might reach its end —all while the object continues its final descent. That strange inversion of time and space is one of the deepest consequences of Einstein's theory.

The idea that time could stretch to infinity isn't just poetic—it's a real prediction of physics. It tells us that black holes aren't just dense objects; they're regions where the structure of reality itself is twisted. To describe what happens beyond the horizon, we'd need equations that merge space and time in ways we don't yet understand. Quantum theory hints that time might not be continuous but granular, built from discrete moments. If that's true, then inside a black hole, time might dissolve into something else entirely—not stopping, but losing its meaning as we know it.

Some physicists believe that time, as a dimension, might even reverse or loop within a black hole. In certain theoretical models, the singularity doesn't mark an end but a transition—a point where space and time flip roles. Inside, movement through space might actually become movement through time, with no way to separate the two. These ideas remain mathematical rather than observational, but they show how far black holes stretch the imagination of physics.

What's most fascinating is that these questions—where matter goes and what happens to time—aren't isolated puzzles. They connect to some of the deepest mysteries about the universe itself. Understanding black holes might help explain how the cosmos began, since the Big Bang shares similar physics in reverse. It might also reveal whether space and time are continuous or built from smaller, indivisible units. In that sense, every black hole might be a natural laboratory for testing the foundations of reality.

WHY "SPAGHETTIFICATION" ISN'T JUST A FUNNY TERM

The term *spaghettification* sounds almost comical, as if it came from a cartoon rather than a physics textbook. Yet behind the lighthearted name lies one of the most dramatic and unsettling predictions of Einstein's theory of gravity. It describes what happens when an object—anything from a spacecraft to a star, or even a single atom—ventures too close to a black hole and gets stretched to its breaking point. The word may sound funny, but the process it describes is anything but. It's one of the clearest examples of just how differently the universe behaves in extreme gravity.

At its core, spaghettification is about **tidal forces**, the difference in gravitational pull across an object's length. We experience tidal forces here on Earth all the time, though they're gentle compared to what happens near a black hole. The Moon pulls slightly harder on the side of Earth facing it than on the far side, creating tides in the oceans. That difference is only a few parts in a million, but it's enough to move entire seas. Now, take that same principle and increase the gravity millions or billions of times. Near a black hole, the difference in pull from head to toe isn't a gentle nudge—it's a violent tear.

Gravity doesn't pull uniformly. The closer you are to a massive object, the stronger the pull. For something falling toward a black hole, the side nearer the hole feels a stronger attraction than the side farther away. If a person were falling feet-first, their legs would feel a stronger pull than their head. That difference grows as they approach the black hole, stretching their body along one axis while compressing it from the sides. The result is a long, thin shape—a cosmic noodle. The process happens to any object, not just humans. Stars, planets, even gas clouds can be elongated by the same effect. It's physics at its most merciless.

This stretching isn't limited to the body as a whole. As the tidal gradient increases, the forces become strong enough to overcome the molecular bonds that hold matter together. At first, it tears apart structures like planets and asteroids. Closer in, it rips apart atoms, then nuclei, until even subatomic particles lose their coherence. Matter stops being matter in any recognizable sense. It's stripped down to fundamental particles and energy, all drawn along the same fatal path inward.

The reason this happens has to do with the steepness of space-time around a black hole. Space and time aren't flat there—they're curved inward, and the slope gets steeper the closer you get. In everyday situations, that curve is so mild we don't notice it. Even near a planet, gravity's change over a few meters is negligible. Near a black hole, though, the curve becomes so sharp that one part of an object can be much deeper in the gravitational field than another. It's as if your feet are sliding down a cliff while your head is still standing on the edge. That difference is what creates the stretching effect.

In smaller black holes—those a few times the mass of the Sun—this process happens violently just outside the event horizon. The gravitational field changes rapidly over short distances, creating tidal forces strong enough to tear apart anything that gets close. In larger, supermassive black holes, the same process happens more gently. The gradient in gravity near the event horizon is much smaller because the hole's size spreads it out. You could, at least in theory, cross the horizon of a supermassive black hole without feeling immediate destruction. The stretching would come later, deeper inside, where the curvature becomes extreme. That's why the scale of the black hole makes a difference in how quickly spaghettification occurs.

Scientists describe this effect using the concept of **tidal acceleration**—the rate at which gravity changes from one point to another. If you were to calculate it for Earth, the difference in pull between your head and feet is practically zero. Near a small black hole, it could be billions of times stronger. The math behind it shows how black holes magnify the normal rules of gravity into something almost unimaginable. These aren't forces that simply push or pull; they distort the very geometry of the universe around them.

Spaghettification also helps illustrate why black holes are not just compact objects but **distortions of space-time** itself. The stretching happens not because something is physically pulling you apart in the usual sense, but because the shape of space-time forces your different parts to follow diverging paths. The lower part of your body is moving along one curve of space, while the upper part is still following another. The difference grows rapidly, stretching you across the contours of gravity's geometry. To an observer watching from afar, the process would appear frozen in time, as if you were being slowly drawn out like taffy, your light fading and reddening until it disappears entirely.

While it might sound theoretical, spaghettification isn't purely abstract. Astronomers have actually witnessed its effects on stars that venture too close to supermassive black holes. When a star gets caught by one, the difference in gravitational pull between the near and far sides can rip the star apart, scattering its gas into long, luminous streams.

Those stellar remnants swirl into an accretion disk, heating up and glowing as they fall inward. These events, known as *tidal disruption events*, are some of the brightest flares in the cosmos, a visible sign of spaghettification on a massive scale.

The reason the term endures—beyond its humorous sound—is because it captures something that's both simple and horrifyingly vivid. It's easy to picture, and it describes a process that has no equal in everyday experience. There's no gradual destruction, no intermediate stage where rescue might be possible. Once the tidal forces reach a certain point, the disassembly of matter is inevitable. Every atom becomes a thread in a long, narrowing stream, drawn toward the core of gravity itself.

What makes spaghettification more than a curiosity is what it tells us about **the limits of the universe's laws.** The fact that space-time can exert such uneven forces that even atomic structure fails shows that gravity, in its most extreme form, can overcome all other forces of nature. Electromagnetism, which binds atoms; the strong nuclear force, which holds protons and neutrons together; the weak force, which governs radioactive decay—all are powerless in the face of gravitational gradients that steep. It's a reminder that black holes aren't just massive—they represent a regime where the hierarchy of physical laws reverses.

There's another layer to this that scientists find fascinating. Spaghettification doesn't just destroy matter—it also affects *how time behaves* for anything falling inward. As you approach the event horizon, not only are you being stretched spatially, but your experience of time also changes. To an external observer, the stretching and fading would take an eternity. To the falling object, it all happens quickly, within moments of crossing the horizon. That blending of spatial and temporal distortion adds depth to what might otherwise sound like a simple mechanical effect. It's not just space being twisted—it's reality itself.

The fact that such an extreme process can be described with straightforward physics is part of what makes it so compelling. The equations that predict spaghettification come directly from Einstein's general theory of relativity. They describe how geodesics—the natural paths through curved space-time—diverge or converge. Where they diverge dramatically, stretching occurs; where they converge, compression follows. Together, those effects explain why objects elongate in one direction and squeeze in another. The math is elegant; the outcome, catastrophic.

Despite its grim implications, spaghettification plays an important role in helping scientists test the boundaries of relativity. Observations of stars torn apart by black holes allow astronomers to measure the strength of gravity, the rate of energy release, and the behavior of matter near the event horizon. Each event becomes a natural experiment, revealing how close reality comes to the predictions of theory. Every stretched star, every flare of falling gas, is a window into the most intense gravitational conditions nature can produce.

5

SEEING THE UNSEEABLE

If black holes don't emit light, how can we possibly know they're there? That question gets to the heart of one of astronomy's most fascinating challenges. A black hole, by definition, hides from direct view. Its gravity is so powerful that not even light can escape, which means telescopes can't see it in the traditional sense. Yet scientists have not only confirmed their existence but also mapped their movements, weighed them, and even captured their silhouettes. The trick lies in looking not *at* black holes, but *around* them—studying how their immense gravity affects everything nearby.

The simplest way to detect something invisible is to watch how it influences what we *can* see. Gravity provides that evidence. Just as wind can't be seen directly but reveals itself by the sway of trees or the curl of waves, black holes show their presence by the way they tug on stars, gas, and light. Their gravitational influence is unmistakable once you know how to recognize it.

One of the earliest and most reliable methods is through **stellar motion**. In some systems, a visible star appears to be orbiting an unseen companion. By carefully tracking the star's movement—how it speeds up, slows down, and shifts position—astronomers can calculate the mass of whatever it's circling. If that mass is far too large to be another star, and yet no light is detectable from it, the explanation almost certainly points to a black hole. The most famous example of this kind of detective work is right at the center of our own galaxy.

In the 1990s, two teams of astronomers began observing a cluster of stars near the heart of the Milky Way. Using infrared telescopes to see through the dust that normally hides that region, they tracked how those stars moved year after year. Some followed incredibly tight, fast orbits, whipping around a single, invisible point in space. When the data were analyzed, it became clear that the unseen object had the mass of about four million Suns packed into a space no larger than our solar system. There was only one possible explanation: a supermassive black hole. That object, now known as Sagittarius A*, has since become one of the most studied black holes in the universe.

Another method relies on **accretion disks**—the swirling rings of gas and dust that gather around black holes. Although the black hole itself emits no light, the material in these disks gets compressed and heated to extreme temperatures as it spirals inward. That friction produces intense radiation, often in the X-ray part of the spectrum. Telescopes designed to detect X-rays, like NASA's Chandra Observatory, can pick up these signals even from millions of light-years away. When astronomers spot an X-ray source that flickers rapidly and seems to come from a tiny region of space, it's a strong indicator that a black hole is feeding nearby.

In some cases, the evidence comes from **binary systems**

—pairs of stars where one member is visible and the other is hidden. The visible star may appear to wobble or lose matter from its outer layers. That matter often forms a glowing disk as it's pulled toward the invisible partner. By studying the light and X-rays from that disk, scientists can determine the mass and behavior of the unseen object. When those numbers exceed what a neutron star (the densest known stellar remnant besides black holes) could be, the case becomes clear.

Black holes also reveal themselves through **jets**—powerful beams of energy that shoot out from their poles. These jets aren't emitted from inside the black hole but from the regions just outside the event horizon, where magnetic fields and rotating plasma interact. They can stretch for thousands or even millions of light-years, glowing brightly in radio waves. By mapping these jets, astronomers can trace their origins back to the black holes that power them. It's an indirect form of observation, but one that has proven incredibly reliable.

Another elegant clue comes from **gravitational lensing**, a phenomenon predicted by Einstein's theory of general relativity. Massive objects like black holes bend the path of light that passes near them. If a black hole sits between us and a distant star, it can distort or even magnify that star's light, creating rings or arcs known as "Einstein rings." By analyzing these distortions, astronomers can infer the black hole's mass and position. In some cases, they've even detected isolated black holes wandering through space by watching how they briefly brighten a background star.

More recently, the discovery of **gravitational waves** has opened a completely new way of detecting black holes. These are ripples in the fabric of space-time, created when massive objects accelerate—such as two black holes orbiting each other before merging. When they collide, the resulting burst of gravitational waves travels outward across the universe, stretching and squeezing space as it passes. Instruments like LIGO and Virgo are sensitive enough to measure these tiny distortions, even though they're smaller than the width of a proton. Each signal they detect tells a story about the masses, distances, and spins of the black holes that produced it.

These gravitational-wave detections have confirmed that black holes not only exist but frequently merge, creating new ones in the process. They've also revealed black holes with unexpected properties—ones heavier or lighter than scientists predicted. Each new signal adds another piece to the puzzle, helping us understand how black holes form, grow, and shape their surroundings.

Sometimes, black holes announce themselves through their sheer *absence* of light. Astronomers looking at the centers of galaxies often notice that stars move as though orbiting something massive, but nothing visible occupies that spot. The speeds and trajectories of those stars reveal the gravitational pull of something enormous and compact —something that can only be a black hole. In this way, astronomers can weigh black holes even when there's no glowing disk or bright jet to mark their presence.

In smaller, quieter cases, scientists turn to subtle methods like **microlensing**—a temporary brightening of distant starlight caused by the gravitational field of a passing black hole. These brief flashes can be recorded with precise instruments, allowing researchers to detect lone black holes drifting through the galaxy. It's a remarkable thought: we can infer the existence of invisible objects that may not be interacting with anything else, simply by watching how they bend light for a fraction of a second.

Each of these techniques—stellar motion, X-ray emission, jets, lensing, and gravitational waves—relies on one central principle: gravity leaves fingerprints. Even if a black hole hides from direct observation, its influence radiates outward in patterns we can measure. In that sense, black holes are like actors on a stage who refuse to step into the spotlight but still move the scenery, shift the lighting, and send ripples through the set. We may never see the actor directly, but we can map every move they make.

This detective work requires not only powerful technology but also patience and creativity. Astronomers must combine data from across the electromagnetic spectrum—radio, infrared, visible light, ultraviolet, X-rays, and gamma rays—to build a full picture. Each wavelength tells a different part of the story. Radio waves trace jets and disks; X-rays reveal high-energy collisions; visible light tracks stars caught in orbit. When those clues are assembled, the invisible object at the center becomes unmistakable.

WATCHING THEIR EFFECTS

The universe doesn't hand out easy clues when it comes to black holes. You can't see one directly, and yet their presence is undeniable once you know what to look for. They shape the movement of nearby stars, twist the paths of light, and sculpt the very space around them. If a black hole were sitting quietly on its own in deep space, we'd have little hope of finding it. But the universe is rarely that tidy. Most black holes have neighbors—stars, clouds of gas, or dust—and their powerful gravity leaves clear signatures in how those neighbors behave.

The first and most dramatic effect shows up in the way **stars move**. A black hole acts like an invisible anchor, holding nearby stars in fast, looping orbits. Astronomers can track those orbits over years or decades, watching stars whip around an unseen point at incredible speeds. It's not guesswork; it's precision measurement. Each motion tells a story about the mass pulling at it. When the mass turns out to be millions or billions of times that of our Sun—and yet emits no light—it can only mean one thing: a black hole.

This method has given us one of the clearest views into our own galaxy's center. Near Sagittarius A*, the supermassive black hole at the heart of the Milky Way, stars race through space in tightly curved paths. Some move so quickly that they complete their orbits in just a few years. To put that in perspective, our Sun takes more than 200 million years to circle the Milky Way once. Watching these inner stars move is like seeing a dance choreographed by gravity itself. They swing close to the center, accelerate sharply, and then swing back out, all controlled by a force that's invisible but measurable.

These observations aren't just about proving a black hole exists—they let scientists calculate details like its mass, size, and even how fast it spins. Every flicker in a star's position or brightness becomes a data point. It's an elegant demonstration of how the universe leaves clues hidden in plain sight. The black hole doesn't need to shine for us to know it's there; the stars around it become its spotlight.

The influence of black holes isn't limited to stars. **Gas and dust** tell an equally compelling story. When a black hole pulls in nearby matter, that material doesn't fall straight in. It swirls, forming an accretion disk where friction and compression heat it to extraordinary temperatures. The gas glows with energy, sometimes brighter than the entire galaxy that surrounds it. In those cases, we call the black hole *active*, and its light can be seen across billions of light-years. Even though the black hole itself remains dark, the environment around it becomes one of the most luminous regions in the cosmos.

This glowing gas gives scientists a window into how black holes grow and interact with their surroundings. By studying the radiation from these accretion disks, astronomers can measure how fast matter is moving, what temperatures it reaches, and how the black hole's magnetic fields shape its behavior. X-rays and gamma rays, which come from the hottest, most energetic parts of the disk, reveal details about how matter behaves just before crossing the event horizon. These aren't just pretty lights; they're signals from the edge of the impossible.

Sometimes, black holes create **jets**—narrow streams of particles that shoot out from their poles at nearly the speed of light. These jets can stretch for thousands or even millions of light-years, cutting through space like cosmic searchlights. The mechanism behind them isn't fully understood, but it involves magnetic fields and the black hole's rotation. What's remarkable is that these jets often carry away more energy than what falls into the black hole itself. They shape entire galaxies, stirring up gas and triggering—or quenching—new star formation. By tracing the direction and strength of these jets, astronomers can learn not just where a black hole is, but also how it's influencing its galactic environment on a vast scale.

Even when black holes aren't feeding dramatically, their presence can still be detected through **subtle gravitational effects.** In binary systems, where a black hole orbits a normal star, the visible star might appear to wobble or vary in brightness as it moves through its orbit. If the star occasionally loses material, that gas can spiral into the black hole, creating a brief flash of X-rays before vanishing. These flares are like the heartbeat of the system, brief bursts that confirm the hidden companion's identity.

Another way scientists observe black holes is through the way they bend **light itself.** Einstein's theory of general relativity predicts that mass curves space, and black holes —being incredibly dense—create the strongest curvature of all. Light passing near one doesn't travel in straight lines; it bends, sometimes looping around before continuing on its way. When a black hole lies between Earth and a distant star, that bending can focus the star's light, making it appear brighter for a short time. This phenomenon, called gravitational lensing, lets astronomers detect black holes that might otherwise remain invisible.

In some spectacular cases, lensing can create multiple images of the same object or even a perfect circle of light known as an Einstein ring. These distortions are more than curiosities; they're measurable effects that allow scientists to calculate the mass and position of the black hole responsible. The way light bends is a kind of cosmic fingerprint—a unique pattern that reveals the shape and strength of the gravitational field.

Black holes also leave their mark on the **motion of entire galaxies.** In many galaxies, stars near the center orbit faster than expected, suggesting an enormous but hidden mass at the core. When astronomers measure the velocity of those stars and compare it to the galaxy's light output, they can estimate how massive the central black hole must be. Over time, this has led to a surprising discovery: nearly every large galaxy seems to host a supermassive black hole at its center, and its mass is linked to the size of the galaxy itself. That relationship suggests a deep connection between black holes and the evolution of galaxies—a partnership that shapes the universe on scales far larger than any single object.

Even **light escaping from around a black hole** carries information about what's happening nearby. As photons climb out of the deep gravitational well, their wavelengths stretch, shifting toward the red end of the spectrum. This "gravitational redshift" allows scientists to map how spacetime curves near the event horizon. By analyzing how much light has been stretched, they can estimate how strong the gravity is and how close the emitting material was when it radiated that light. Every spectrum, every color shift, becomes a way to probe an environment that no probe or spacecraft could ever reach.

Astronomers also study **variability**—the way light from the region around a black hole flickers and changes. These fluctuations can happen in seconds or over years, depending on the size of the black hole and the amount of matter falling into it. They reveal not only the physics of accretion but also the geometry of the system—whether the disk is stable, tilted, or warped by the black hole's spin. In this way, even the rhythm of light becomes a diagnostic tool.

GRAVITATIONAL WAVES

Every part of the universe moves, stretches, and shifts. Stars explode, galaxies collide, and black holes spiral together in an endless dance of gravity. For a long time, scientists believed these colossal motions might create ripples—disturbances that travel through space itself, carrying traces of the events that caused them. These ripples are called *gravitational waves*, and they are one of the most extraordinary ways we've learned to study black holes.

Gravitational waves aren't made of light or sound.

They're literal distortions in space-time—the invisible grid that underlies everything in the universe. When a massive object accelerates, it bends and stretches this fabric, sending out waves that ripple across the cosmos at the speed of light. Think of tossing a heavy stone into a calm lake: the splash sends circular waves outward, but instead of water, these waves travel through the structure of space itself.

Black holes are among the strongest sources of gravitational waves. When two of them orbit each other, their combined gravity twists space into a kind of whirlpool. As they spiral closer, they move faster and faster, releasing energy in the form of gravitational waves. When they finally merge, the collision sends out a burst powerful enough to shake the universe. The waves that reach Earth are incredibly faint by the time they arrive—stretching and squeezing space by less than the width of a proton—but with precise instruments, scientists can detect them.

That detection was once only a dream. The concept came from Einstein's equations over a century ago, but even he wasn't sure they could ever be measured. The distortions are so small that they seemed beyond human reach. Yet in 2015, that changed. Two black holes merged over a billion light-years away, and the faint tremor of their collision finally reached Earth. For the first time, the detectors built to catch such a signal—part of a system called LIGO—recorded the tiny oscillation. It was a historic moment: the first direct evidence that gravitational waves exist and that black holes not only form but collide and merge.

The signal lasted less than a second, but it told an enormous story. By studying the shape and strength of the wave, scientists could tell that the two black holes were about 30 times the mass of the Sun each, and that when they merged, they created a new black hole slightly smaller than the sum of its parts. The missing mass had been converted into pure energy—energy that traveled across the universe as the very waves being measured. In that brief instant, more energy was released than all the stars in the visible universe combined.

The fact that a faint ripple from such a distant event could be detected on Earth is astonishing. LIGO's instruments use laser beams bouncing between mirrors placed kilometers apart. As a gravitational wave passes through, it stretches one arm of the detector while compressing the other by an infinitesimal amount. That change alters the pattern of light in a way scientists can measure. The setup is so sensitive that even the vibrations from a passing truck or a nearby thunderstorm can interfere with the readings. To confirm a true signal, multiple detectors at different locations must record the same wave at nearly the same time.

Once that first detection was confirmed, the field of gravitational wave astronomy was born. Each new signal adds a layer to our understanding of black holes. By comparing the waveforms—the patterns of rise and fall in the data—scientists can reconstruct what happened during the merger: how massive the black holes were, how fast they spun, and how their final collision unfolded. It's like listening to an orchestra where each note corresponds to a cosmic event. The louder and lower the tone, the more massive the black holes. The quicker the oscillations, the closer they were before merging.

One of the most remarkable aspects of gravitational waves is that they carry information light can't. Traditional telescopes rely on electromagnetic radiation—visible light, radio waves, X-rays—to study the universe. But those forms of light can be blocked, scattered, or absorbed by gas and dust. Gravitational waves, on the other hand, travel through everything untouched. They give us a direct connection to events hidden from ordinary sight. When two black holes merge, no light escapes to tell the story—but the ripples in space do.

This ability to "listen" to the universe rather than just "look" at it has opened a new kind of observation. Before 2015, everything known about black holes came from how they affected their surroundings—how stars moved around them or how gas glowed as it fell inward. Now, scientists can detect black holes that exist in isolation, far from any nearby matter. Even if they emit no light or radiation, their mergers produce gravitational waves that announce their existence across billions of years.

Gravitational wave data also helps solve deeper mysteries. For example, scientists can now estimate how often black holes merge, how massive they typically are, and how they might have formed. Some may be the remnants of massive stars collapsing at the end of their lives. Others could be ancient black holes that formed when the universe was still young. The differences in their signals provide clues about their origins—whether they were born together as a pair or met later through chance encounters in dense star clusters.

In a way, gravitational waves act as a kind of cosmic seismograph. Just as earthquakes send vibrations through the Earth that reveal its inner structure, these waves carry information about the dynamics of space-time itself. They can show how strong gravity behaves in extreme environments, where conventional theories break down. Every detection is a test of Einstein's general relativity under conditions too intense to reproduce on Earth. So far, his equations have held up perfectly, but each new observation pushes them to their limits.

There's also something poetic about how gravitational waves let us experience the universe through time rather than light. Each wave that passes through Earth slightly stretches and compresses everything—our planet, our instruments, even our bodies—though by an amount too small to feel. In a sense, every detection is a handshake across cosmic distance, a whisper from an event that ended long before life on Earth began.

The information they bring isn't limited to black holes. Neutron stars—the ultra-dense remnants of exploded stars—also produce gravitational waves when they collide. In 2017, scientists detected such a merger and, for the first time, matched the wave signal with light from telescopes around the world. That event confirmed that these collisions create heavy elements like gold and platinum. It was a glimpse into how the ingredients of planets and life itself are forged in cosmic violence.

For black holes, gravitational waves remain our most direct messenger. Unlike light, which comes from the matter around them, these waves arise from their very motion. They reveal black holes as dynamic, evolving systems—not static voids but participants in an ever-changing universe. As detectors improve, we're beginning to hear not just the biggest, loudest events, but subtler ones too: smaller black holes merging, or maybe even intermediate-sized ones that bridge the gap between stellar and supermassive varieties.

THE FIRST PHOTOGRAPH

For decades, black holes were thought to be forever beyond our view—objects we could prove existed only through their effects, never by direct sight. Then, in April 2019, that barrier broke. Astronomers unveiled the first-ever photograph of a black hole: a glowing ring surrounding a dark core, taken by the Event Horizon Telescope (EHT). It was more than an image; it was a turning point in how we see the universe. The subject of that photo lay in the heart of a distant galaxy called Messier 87, or M87 for short, a colossal elliptical galaxy about 55 million light-years away.

The M87 black hole isn't a small or subtle presence. It's a monster—about 6.5 billion times the mass of the Sun, large enough that our entire solar system could fit comfortably within its event horizon. The galaxy that hosts it had long been known to shoot a narrow jet of high-energy particles thousands of light-years into space. Scientists suspected for decades that only a supermassive black hole could power such a jet, but until the EHT's work, no telescope had ever seen the structure that proved it.

Capturing that first image wasn't simply a matter of pointing a powerful telescope at the target. The challenge was that even though the M87 black hole is unimaginably huge, it's also unimaginably far away. From Earth, it appears about the same size as a doughnut sitting on the Moon. No single telescope on the planet could achieve that level of detail. The solution required a different kind of thinking—a way to turn the entire Earth into one enormous telescope.

The Event Horizon Telescope isn't one device but a global network of observatories spread across continents—from Hawaii to Spain, Chile to the South Pole. Each telescope observes the same target at precisely the same time, recording radio waves with atomic-level accuracy. By combining all these observations, scientists can create the effect of one telescope as large as the planet itself. This technique, called *very long baseline interferometry*, allows astronomers to resolve details that would otherwise be impossible to see.

The coordination behind this project was staggering. The data collected from each observatory couldn't simply be emailed or uploaded—it was too vast. Instead, hard drives containing petabytes of information had to be physically flown to central processing centers, where specialized supercomputers combined and compared the data. Every piece had to align perfectly in time and position, down to billionths of a second, to recreate a single, coherent image.

When the team finally assembled the data, what emerged was a bright ring of light with a deep shadow in its center. The ring came from superheated gas swirling around the black hole's event horizon, glowing as it was torn apart by gravity and friction. The dark region wasn't the black hole itself—it was the shadow cast by its gravitational pull, the zone where light could no longer escape. The symmetry and clarity of that shape were exactly what Einstein's theory of general relativity predicted.

The image didn't come out of nowhere. It was the result of decades of preparation—technological advances, global cooperation, and sheer persistence. Many scientists who worked on it had dedicated their entire careers to the goal of seeing a black hole's shadow. When the photograph was released, it wasn't just a scientific triumph; it was a human one. The collective effort of over 200 researchers from multiple countries came together to produce something no one had ever witnessed before: direct visual evidence of a black hole.

At first glance, the picture may seem simple—a glowing orange ring against a black background. But it's one of the most data-rich images ever created. Every pixel represents light that has traveled across millions of years, bent and distorted by gravity on its way to Earth. The ring's brightness isn't uniform; it's brighter on one side because the black hole is spinning. The light from gas moving toward us gets boosted in intensity by the Doppler effect, while light from gas moving away is dimmed. This uneven glow confirms that the black hole in M87 rotates, dragging space itself along with it.

Beyond its beauty, the image gave scientists a way to measure the black hole's mass and test Einstein's equations in extreme conditions. The size of the dark region matched the predictions nearly perfectly, reinforcing our understanding of how space-time behaves under intense gravity. It also offered insight into the connection between black holes and the galaxies they inhabit. The energy released by material swirling around the M87 black hole fuels the galaxy's powerful jet, which stretches across intergalactic space. Seeing the source of that jet linked theory to observation in a way that had never been achieved before.

The public reaction to the image was immediate and profound. For many, it transformed black holes from abstract math into something tangible. Before that, they were the stuff of theory and metaphor—enigmas hidden behind equations. Now there was a photograph, proof that these cosmic voids weren't just imagination or speculation. They were real, observable, and part of the same universe we inhabit. The image became a symbol of curiosity and collaboration, a reminder that humanity can capture even the most elusive phenomena when it works together.

The photograph also marked a new chapter in astronomy—one where we don't just infer black holes through their effects, but *see* their structure. For decades, scientists had looked at stars whipping around invisible centers or at high-energy radiation from accreting gas and said, "There must be a black hole there." Now they could point to an actual image and say, "There it is." That shift from inference to observation has enormous implications. It bridges the gap between theory and experience, giving us a clearer window into the extreme physics near an event horizon.

The M87 image also provided a foundation for future discoveries. Just three years later, in 2022, the EHT team released another historic image—this time of Sagittarius A*, the black hole at the center of our own Milky Way. That one was even more difficult to capture because Sagittarius A* changes rapidly; its accretion disk shifts on timescales of minutes, whereas M87's changes over days or weeks. Yet the same principles applied: a glowing ring, a dark shadow, and an exact match to relativity's predictions.

The success of the M87 image has inspired scientists to refine their methods even further. Plans are underway to expand the Event Horizon Telescope network with more observatories, improving resolution and enabling video-like sequences that could show matter orbiting in real time. In the future, we might not just see a still frame of a black hole but watch the movement of light itself swirling around the edge of space and time.

6

TIME, SPACE, AND THE WEIRD PHYSICS AROUND BLACK HOLES

Before Einstein, the universe seemed simple enough. Space was thought to be a kind of empty stage where planets, stars, and galaxies moved, and time ticked along steadily like a universal clock, the same for everyone everywhere. Gravity, as described by Isaac Newton, was a force—something that pulled objects together. It worked beautifully for centuries. Newton's equations could predict how an apple fell, how the Moon orbited Earth, even how comets curved through the solar system.

But there were cracks in that picture—tiny, stubborn discrepancies that Newton's laws couldn't quite explain. The orbit of Mercury, for example, didn't behave exactly as expected. And as scientists learned more about light, they found that it didn't seem to fit neatly into Newton's framework. These were small problems, but they hinted at something deeper, something that would eventually require a complete rethinking of how the universe works.

Einstein's theory of relativity didn't just patch up the old model—it replaced it with something profoundly different. He showed that space and time aren't separate, independent things. They're woven together into a single entity called *space-time*, a kind of four-dimensional fabric that can stretch, bend, and curve. What we perceive as gravity, he said, is really just the result of that fabric curving around massive objects.

It's a bold idea, but it explains everything Newton's gravity could and much more. In Einstein's view, Earth orbits the Sun not because the Sun "pulls" it, but because the Sun's mass bends the space around it, and Earth simply moves along the curve of that distortion. The same way a marble rolls around a dent in a stretched sheet, planets move through the curved geometry created by stars and other massive bodies.

This new understanding of gravity led to predictions that sounded outlandish at first. Light, for example, should bend when passing near a massive object, even though light has no mass. Time itself should slow down in strong gravitational fields. These weren't philosophical musings—they were testable consequences of his equations. And over the years, every one of them has been confirmed through observation and experiment.

The shift Einstein introduced wasn't just mathematical —it was conceptual. He forced humanity to abandon the comforting idea of a fixed, absolute universe. In Newton's world, space and time were like a grid laid over reality: predictable, consistent, unchanging. In Einstein's universe, they are dynamic, responsive, alive. When a star moves, it disturbs space-time; when a planet orbits, it leaves ripples behind. Nothing exists in isolation. Every mass affects every other through the geometry of space itself.

This interconnection has consequences that ripple through all of modern physics. One of the most striking is **time dilation**—the fact that time doesn't pass at the same rate everywhere. Near a strong source of gravity, time slows down. At high speeds, it stretches differently for moving observers. Einstein unified space and time into one system, which means if space can bend, time must bend with it.

To most of us, this feels completely alien. Our everyday experience tells us time is constant. But we've verified Einstein's prediction countless times. Astronauts orbiting Earth age ever so slightly more slowly than people on the ground. Atomic clocks flown on airplanes tick at different rates than those that stay at sea level. Even the GPS systems we rely on for navigation wouldn't function without accounting for both gravity and motion affecting time. What sounds like science fiction turns out to be necessary for something as ordinary as finding the nearest gas station.

Relativity also revealed that matter and energy are two sides of the same coin. That insight, captured in the famous equation $E = mc^2$, tells us that mass can be converted into energy and vice versa. It explains how the Sun shines—nuclear fusion in its core turns matter into light and heat. It also paved the way for understanding nuclear reactions, both constructive and destructive. That single equation reshaped not just physics but technology, showing that the universe's most basic ingredients are interchangeable expressions of the same thing.

When Einstein first proposed these ideas, they were met with skepticism. They contradicted centuries of intuition and seemed absurdly abstract. But his equations had a remarkable quality: they predicted phenomena that could be checked. One of the earliest tests came in 1919, when astronomers observed starlight bending around the Sun during an eclipse, exactly as relativity predicted. That moment turned Einstein from an obscure physicist into a global figure and confirmed that the universe truly followed his strange, elegant rules.

The effects of relativity become most dramatic under extreme conditions—precisely the kind found near black holes. Those objects are the ultimate expression of curved space-time, where Einstein's equations reach their limits. In a sense, black holes are nature's way of showing what happens when mass pushes space-time to the breaking point. Without relativity, they wouldn't even make sense as concepts. Newton's laws can't describe a region where time slows to a crawl or light bends back on itself. But Einstein's framework handles it perfectly—up to a point. Beyond the singularity, even relativity gives way, hinting at deeper physics still waiting to be discovered.

Relativity also changed how scientists think about the universe as a whole. When Einstein applied his equations to the cosmos, he found that space-time itself must either expand or contract—it couldn't remain static. That realization, confirmed decades later with the discovery of the expanding universe, laid the groundwork for the Big Bang theory. It showed that space-time is not a backdrop but an evolving entity, one that can stretch over billions of years. Every galaxy, every cluster of stars, moves through this shifting geometry like dust drifting through a vast, elastic sea.

Perhaps the most surprising legacy of Einstein's work is how it connects the smallest and largest scales of existence. His theory explains the motion of planets and the bending of light around stars, but it also influences quantum physics, cosmology, and even modern technology. It bridges the human-scale world of experience with the cosmic ballet of galaxies and the microscopic jitter of subatomic particles.

There's also a philosophical dimension to what relativity introduced. It changed not only how scientists describe reality, but how we think about our place in it. Time isn't universal, meaning there's no single "now" that applies everywhere. Two events that seem simultaneous to one observer might not be to another. That means the universe doesn't operate on an objective, synchronized schedule—it's more like a collection of personal timelines woven together by the geometry of space-time.

What's remarkable is how something so abstract ended up explaining everything from the behavior of light to the shape of the cosmos. Einstein's equations have guided spacecraft navigation, predicted black holes and gravitational waves, and allowed the creation of technologies that depend on precision timekeeping. Yet their beauty lies in their simplicity. They reduce the complexity of the universe to a few concise ideas: that matter tells space how to curve, and curved space tells matter how to move.

It's hard to overstate how radical this was. Before Einstein, space and time were thought to be eternal, unchanging stages. After Einstein, they became dynamic, participatory parts of the play. The actors—the planets, stars, and galaxies—no longer move across a flat stage; their motion reshapes the very surface beneath their feet.

TIME DILATION

Time feels like one of the most steady, reliable things in life. Clocks tick forward at the same pace, the Sun rises and sets every twenty-four hours, and no matter where you go on Earth, a second is a second. But near a black hole, that constancy falls apart. The closer something gets to the immense gravity of one of these cosmic giants, the slower time itself runs. To an observer far away, it's as if time almost stops at the edge.

This strange effect—*time dilation*—is one of the most mind-bending consequences of Einstein's theory of relativity. It happens because gravity doesn't just pull on matter; it stretches and bends space-time itself. And when space-time bends, it affects how time flows. Near a black hole, where gravity reaches its most extreme form, this distortion becomes profound.

At first glance, it might sound like science fiction. But time dilation is not an abstract idea or a hypothetical theory —it's a measurable, proven reality. Even on Earth, we've seen it happen. Astronauts orbiting in space experience slightly slower time than people on the ground. The difference is tiny—fractions of a second—but measurable with atomic clocks. Near a black hole, that difference wouldn't be tiny at all; it would be enormous.

To understand why, it helps to think about how gravity affects the passage of time. In Einstein's universe, space and time are intertwined into one continuous fabric—space-time. A massive object like a star or planet creates a curve in this fabric. The deeper the curve, the stronger the gravitational pull, and the slower time runs. For small masses like Earth, the effect is barely noticeable. But for a black hole, the curve becomes so steep it's almost vertical. Time at its surface, the event horizon, slows to an extreme degree compared to anywhere else.

Let's break that down more concretely. Suppose an astronaut is hovering near a black hole while another remains at a safe distance. For the astronaut near the black hole, everything feels normal. Their heart beats at its usual rhythm, their watch ticks at a steady pace, and nothing seems unusual. But to the distant observer watching through a telescope, the story looks very different. They would see their friend's movements slow dramatically as they got closer to the event horizon. Each second of the astronaut's time might stretch into hours, years, or even centuries from the outside perspective.

The reason for this difference is that gravity affects not just physical motion but the very structure of time. The stronger the gravitational field, the more energy it takes for light—and therefore information—to escape it. Light leaving a region near a black hole gets stretched, losing energy and shifting to longer wavelengths. This same stretching applies to every process that unfolds there, including the ticking of a clock. Time itself appears to lose momentum.

If the astronaut were to hover just outside the event horizon, they could, in principle, watch the entire future of the universe unfold in fast-forward while experiencing only a few minutes of their own time. Distant galaxies would age, stars would die, and cosmic structures would shift while the astronaut's personal clock barely moved. It's not a trick of perception; it's a fundamental difference in how time flows in two places experiencing different gravitational strengths.

Now, if the astronaut crossed the event horizon, things become even more extreme. From the outside, it would appear as if they never quite crossed it. Their motion would seem to slow endlessly, their light fading and reddening until they vanished. But from their own perspective, the crossing would happen smoothly, in a finite amount of time. Both experiences are valid within their own frames of reference, even though they contradict each other when described together. That's part of what makes black holes—and relativity in general—so hard to wrap our heads around.

What's remarkable about this phenomenon is that it's not limited to black holes. The same effect exists around any massive body, just to a much smaller extent. For example, time passes slightly faster at the top of a mountain than at sea level because gravity is weaker higher up. That means if you spent your entire life living on the top floor of a skyscraper, you'd technically age a few nanoseconds faster than someone on the ground floor. It's imperceptible to us but essential for technologies that depend on ultra-precise timing, such as GPS satellites. Those satellites orbit Earth in weaker gravity, and if engineers didn't adjust for time dilation, their calculations would quickly drift off course.

What black holes do is take that subtle, everyday effect and amplify it beyond anything we can experience elsewhere. The warping of space-time near them is so intense that it transforms how cause and effect play out. From a distance, light escaping a black hole's pull seems to take longer and longer to travel. Each photon climbs uphill through space-time, losing energy as it goes, which stretches its wavelength. That's why scientists say the region around a black hole's edge appears "redshifted." The deeper you go into that well of gravity, the slower everything gets, and the more energy light must expend to escape.

There's also a fascinating link between time dilation and energy. Einstein showed that gravity and acceleration are two sides of the same coin. If you were in a rocket ship accelerating upward at a constant rate, time for you would slow relative to someone at rest. That's because acceleration creates a kind of artificial gravity. In both cases, whether you're near a massive object or moving rapidly, space-time itself dictates the flow of time. Black holes simply represent the most extreme example of this rule.

The practical implications stretch beyond the theoretical. Time dilation changes how astronomers interpret what they see. When they study the regions close to black holes, they must account for the fact that processes unfolding there—like gas swirling in the accretion disk—appear slower than they really are. Light emitted from that material also carries the fingerprints of time dilation, encoded in how its spectrum shifts. Understanding those effects allows scientists to calculate the black hole's mass, spin, and even how close matter can orbit before it falls in.

There's another subtlety to this: from the inside of a strong gravitational field, the rest of the universe seems to speed up. The astronaut near the black hole might glance outward and see the stars flickering and moving faster than usual. That's because while time is slower for them, it continues normally elsewhere. It's a two-way distortion, and it means that "now" becomes a flexible concept rather than a shared moment. Each observer carries their own timeline, shaped by their environment's gravity.

What makes time dilation so fascinating is how it challenges the basic human sense of time as universal. We're used to thinking of seconds and minutes as fixed units, but near a black hole, they become elastic. The closer something gets to the event horizon, the more stretched each moment becomes. Eventually, the difference between "before" and "after" nearly disappears at the edge, where gravity and time reach a standstill relative to the outside universe.

SPACE GETS STRETCHED

Gravity is one of those things we grow up accepting without much thought. Drop a ball, and it falls. Planets circle the Sun. We feel heavy on the ground because Earth pulls us down. For centuries, that explanation worked well enough —gravity as a kind of invisible force reaching across space. But Einstein changed that view completely. He showed that what we call gravity isn't a pull at all. It's the shape of space itself.

That shift—from force to curvature—was one of the biggest leaps in human understanding. It took an idea that seemed almost mystical and turned it into geometry. To grasp what that means, it helps to think of space not as empty, but as something flexible, like a sheet or fabric. Matter, whether it's a planet or a star, doesn't sit on this fabric passively. It bends it. And once it's bent, other objects move according to that shape. What we experience as gravity—the tendency for things to fall toward one another —is really just the result of them following curved paths through distorted space.

This concept, called the *curvature of space-time*, is at the heart of relativity. It's what explains everything from falling apples to the orbits of planets and even the structure of the universe itself. But to see why it matters, it helps to start with something familiar.

If you've ever watched water flow around a rock in a stream, you've seen how motion adapts to shape. The water doesn't need to know where to go; it simply follows the contours of what's underneath. The rock doesn't "pull" the water—it changes the path the water can take. Space works in much the same way. Massive objects like stars or planets reshape the geometry around them. Smaller objects—

whether they're marbles or moons—move along those curves naturally. They're not being tugged by an invisible string; they're just following the simplest possible path through the warped landscape.

This kind of path is called a *geodesic*. In flat space, a geodesic is a straight line. In curved space, it's the shortest route between two points, even if it doesn't look straight to us. Airplanes flying over the Earth's surface, for instance, take curved routes that are actually the shortest possible paths on a sphere. The same principle applies in the universe: what we call "orbiting" is simply following a curved geodesic through space shaped by gravity.

Now, the interesting thing about curvature is that it affects both space and time together. Einstein showed that mass doesn't just warp space—it warps the entire space-time fabric. That means when a planet like Earth moves around the Sun, it's not circling because the Sun's gravity pulls it. It's gliding along a bent path in the four-dimensional fabric created by the Sun's mass. If the Sun suddenly vanished (which it can't, of course), Earth wouldn't be yanked outward instantly—it would continue along the straightened path of space-time that had been curved until that moment.

This view explains some of the strange behaviors that older theories couldn't. Take Mercury's orbit, for example. For years, astronomers noticed that its path shifted slightly more than Newton's equations predicted. Einstein's curvature model explained that extra wobble perfectly. The Sun's mass distorts the space-time around it just enough to change Mercury's trajectory, and that tiny effect accumulates over time.

The stretching of space isn't only about movement—it's also about how distance itself behaves. In normal conditions, we think of a meter as a meter everywhere. But in the universe's curved geometry, distance can stretch or contract depending on the gravity involved. Near a black hole, that stretching becomes extreme. Space isn't just distorted; it's stretched to the point that the meaning of "distance" starts to break down. Two points that seem close in coordinate terms might be separated by vast amounts of warped space.

This is what makes black holes so unique. Their gravity doesn't merely hold things—it alters the geometry around them so dramatically that even light, which always moves at the same speed, can't escape. It's not that light is pulled inward faster; it's that the paths it could follow are all curved back toward the center. The structure of space itself directs everything inward. In a sense, the black hole doesn't trap light—it gives it nowhere else to go.

To understand how curvature works, scientists use a mathematical concept called the *metric*. It describes how to measure distances in curved space. On a flat surface, the shortest distance between two points is a straight line, and we can calculate it with simple geometry. But when the surface bends, those rules change. The metric tells us how to adjust for that bending. Einstein's genius was realizing that this same principle could be applied not only to surfaces but to the entire universe, combining space and time into a single structure that reacts to mass and energy.

One way to think of this relationship is through energy balance. Mass and energy tell space-time how to curve; the curvature of space-time tells mass and energy how to move. It's a feedback loop built into the structure of reality. The more mass you have, the deeper the curvature. The deeper the curvature, the more strongly it influences everything nearby.

In everyday life, these effects are mild. The curvature

caused by Earth's mass is enough to keep the Moon in orbit but not enough to distort the geometry in ways we can notice without precise instruments. Yet even here, the principles hold. The tides, for instance, are a result of Earth's oceans responding to the curvature created by both Earth's and the Moon's combined gravity. The water isn't being "sucked" upward—it's following the natural geometry of space shaped by those masses.

Farther out in the universe, where the forces are stronger and scales larger, the stretching of space becomes dramatic. Around neutron stars and black holes, the curvature is so steep that a small change in position can mean an enormous change in the behavior of time and motion. Light bends, trajectories twist, and the concept of "straight" becomes meaningless.

HOW BLACK HOLES LET SCIENTISTS TEST THE LIMITS OF PHYSICS

Black holes are often described as the most mysterious objects in the universe, but for scientists, they're also the most revealing. They're places where the known laws of physics are stretched to their breaking point. Every observation, every equation tested near one, is an opportunity to ask the universe a harder question: *Does our understanding still hold up here?*

Most of modern physics rests on two powerful frameworks. One is **Einstein's theory of general relativity**, which describes how gravity shapes space and time on large scales—planets, stars, galaxies, and the universe as a whole. The other is **quantum mechanics**, which governs the strange behavior of particles at the smallest scales—atoms, electrons, and energy itself. Both theories work beautifully in their own domains. But they don't get along with each other. When scientists try to apply both at once—to places where gravity is incredibly strong and space is incredibly small—the math breaks down.

Black holes are precisely those places. They sit at the crossroads between relativity and quantum mechanics. The gravity around them is so intense that space-time folds in on itself, and yet the regions near their edges are filled with high-energy particles behaving according to quantum rules. If there's anywhere the universe will show us where our understanding fails—or how it might be unified—it's there.

One of the reasons black holes are such effective testing grounds is that they amplify physical effects that are barely noticeable elsewhere. For example, time dilation—the slowing of time in strong gravity—is almost imperceptible near Earth but becomes extreme around black holes. Light bending, another prediction of relativity, is magnified there until it becomes dramatic. These environments don't just confirm Einstein's equations; they push them to the edge, testing whether those equations can survive the most extreme conditions imaginable.

When scientists observe how stars orbit an invisible point in space, as they do around the center of our galaxy, they're watching relativity at work. The precision of those orbits tells us how mass and space interact. If the theory were wrong, those stars wouldn't move the way they do. And yet, every measurement lines up with Einstein's predictions—down to astonishing accuracy. That tells us relativity remains robust, even where gravity is fierce.

But not every question finds a neat answer. Inside a black hole, at the point called the singularity, general relativity predicts that density becomes infinite and space-time curves infinitely inward. That's not a physical description—

it's a sign that the math has reached its limit. Infinity isn't something nature can display; it's what happens when our equations stop being able to describe what's really there. That means there must be a deeper theory waiting to be discovered, one that can handle both gravity and quantum mechanics at once.

In a way, black holes are cosmic laboratories for discovering that deeper theory. Because they sit at the intersection of the two greatest pillars of modern physics, they expose the cracks between them. Relativity describes gravity as smooth and continuous, while quantum mechanics describes everything as discrete and jittery. Reconciling those two pictures—making them fit together—is one of the greatest challenges in science.

Black holes help by giving physicists a place to test where the transition happens. For instance, near the event horizon, where light struggles to escape, the behavior of particles and radiation can reveal how quantum effects influence curved space-time. The famous **Hawking radiation**, proposed by Stephen Hawking, arises from exactly that interplay. According to his idea, quantum effects near the event horizon cause black holes to emit faint radiation and lose mass over time. That's something relativity alone can't explain—it's a purely quantum effect happening in a gravitational field.

Hawking's insight was monumental because it proved that black holes aren't just gravitational traps—they're thermodynamic systems. They have temperature, entropy, and even a kind of "surface area" related to their event horizon. This was the first bridge between quantum physics, thermodynamics, and relativity, showing that all three might be linked in ways we're only beginning to grasp.

The more scientists study these ideas, the more black

holes seem to act like doorways into the underlying structure of reality. They've forced physicists to rethink fundamental ideas about information, energy, and even the nature of space itself. One of the biggest puzzles is known as the **information paradox**—a conflict between quantum mechanics and relativity that black holes bring to the surface.

In quantum mechanics, information can never be destroyed. The state of any system, no matter how chaotic or transformed, is preserved in some way. But in relativity, when something falls into a black hole, it's gone from the universe's observable region forever. That creates a contradiction. If a black hole eventually evaporates due to Hawking radiation, where does the information about what fell in go? Does it vanish, violating quantum theory? Or is it somehow preserved in a form we don't yet understand?

Physicists have spent decades wrestling with that question, and it has driven entire fields of theoretical research. Some believe information might be encoded on the event horizon itself—a concept that led to the **holographic principle**, the idea that everything inside a region of space could be described by data stored on its boundary. Others think that quantum entanglement may hold the key, binding what happens inside and outside a black hole in ways that preserve information across the horizon. Whatever the answer, it's clear that black holes aren't just cosmic curiosities—they're the best tools we have for probing the foundations of reality.

Observations of black holes also let scientists test theories that go beyond Einstein. Alternative models of gravity, dark matter, and cosmic structure can be compared against what we actually see near these objects. For example, when astronomers measured the orbits of stars around Sagittarius

A*, the supermassive black hole at the Milky Way's center, they weren't just proving it existed—they were confirming that space behaves exactly as relativity predicts. If those stars had deviated from the expected paths, it might have suggested a flaw in Einstein's model or pointed to new physics altogether.

Then came one of the most remarkable achievements in modern science: the detection of gravitational waves. When two black holes collide, they don't just merge—they send ripples through the fabric of space-time itself. Those ripples were first detected in 2015 by LIGO, confirming yet another of Einstein's century-old predictions. Each new detection since then provides a fresh test. By comparing the shape and timing of those waves to theoretical models, physicists can see whether relativity continues to hold true under the most violent conditions in the cosmos. So far, it has.

Yet even these successes raise deeper questions. General relativity works beautifully until it meets quantum reality, and quantum theory works perfectly until it meets gravity. Somewhere between them lies a unifying principle—a "theory of everything." Black holes, by their very nature, sit right on that boundary. Every bit of progress in understanding them takes us a step closer to discovering that missing link.

THE "RUBBER SHEET" ANALOGY

When people first learn about how gravity works in Einstein's theory of relativity, they're often introduced to one of the most famous teaching tools in physics: the rubber sheet analogy. The idea is simple and visual. Picture a stretchy rubber sheet pulled tight, representing space itself. Place a heavy ball in the middle of that sheet, and it sinks downward, creating a dip. Then roll a smaller ball across the surface. Instead of moving in a straight line, it curves around the dip, spiraling inward or orbiting if the conditions are right. That curved path is meant to represent how a planet moves around a star, or how light bends near something massive.

This image works remarkably well as a first step toward understanding space-time curvature. It gives shape to an idea that's hard to grasp intuitively: that gravity isn't a force pulling through space, but a change in the shape of space itself. When we look at a planet orbiting a star, the old Newtonian picture says the star is tugging on the planet with an invisible hand. The rubber sheet analogy shows something subtler—the planet isn't being pulled at all; it's simply following the path that space tells it to take.

That's the useful part. It helps people move away from thinking of gravity as an external force and toward seeing it as geometry. The heavier the object, the deeper the dip in the sheet. Smaller objects slide toward the low point not because they're being pulled, but because the ground beneath them is curved. It's a good starting point for understanding why, in relativity, motion and gravity are two sides of the same coin.

But while this image helps build intuition, it's also incomplete—and, in some ways, misleading. The more you unpack it, the more you realize that real space-time isn't much like a rubber sheet at all. The analogy bends in useful ways but breaks in others, and understanding those limits is just as important as the image itself.

The first big limitation is dimensionality. The rubber sheet shows a two-dimensional surface bending into a third dimension. Real space-time has three spatial dimensions and one time dimension. When space-time "curves," it isn't dipping into some higher space we can visualize—it's curving within itself. The rubber sheet relies on the visual cue of depth, but in reality, the curvature of space doesn't require an external direction to bend into. It's more about how distances and paths behave inside the space than about any visible sagging.

In the analogy, the ball sinks because gravity pulls it down into the sheet. But that's a circular explanation—gravity is being used to demonstrate gravity. In the real universe, there's no external downward pull causing mass to make dents in space. The "dip" in space-time isn't formed by something pushing it from above or below. It's an intrinsic property of how mass and energy alter geometry. Space-time curves because the equations of relativity say that's the only way to balance the presence of energy and motion—not because it's being pushed around by another force.

Another issue with the analogy is how it handles—or rather, fails to handle—time. The rubber sheet focuses entirely on the spatial side of curvature. But gravity doesn't just affect space; it affects time, too. A clock near a massive object runs slower than one far away. That time distortion is a fundamental part of what makes gravity work in Einstein's theory. The curved paths we see, like orbits, are not just objects sliding through space—they're following the easiest route through a combined fabric of space and time. The rubber sheet ignores that entirely, flattening four dimensions into two and leaving out one of the most crucial ingredients.

Still, despite these flaws, the analogy does something important: it gives us a way to see curvature as a property of space rather than an invisible pulling force. It lets people connect a physical picture with an abstract concept. Even professional physicists sometimes use it informally when explaining their work, knowing that it's not the whole story but that it gets people pointed in the right direction.

To make the analogy more accurate, it helps to think about what the rubber sheet is *trying* to capture rather than how literally it maps to the real thing. The main message is that mass changes geometry. If you were to draw a grid on the sheet before dropping the heavy ball, those lines would distort and curve around the depression. That's what happens to space-time near a massive body—distances and directions change. Light, which always follows the straightest possible line, now travels along what looks curved to an outside observer.

One way to extend the analogy mentally is to think not about objects rolling across the surface, but about how the grid itself behaves. A circle drawn near the dip on the sheet would have a circumference smaller than you'd expect for its radius if you were using ordinary geometry. That mismatch is one way to describe curvature mathematically, and it's one of the reasons Einstein's equations are expressed in terms of how distances deviate from flatness. In that sense, the rubber sheet is showing us a hint of what curvature means: it's a change in how the rules of geometry apply locally.

Another reason the analogy breaks down is that it makes gravity look like a kind of funnel that always pulls things inward. That's partly true, but not entirely. In reality, space-time curvature can also push things apart. On the largest scales, the universe's expansion is a form of curvature, one that causes distant galaxies to move away from each other. Black holes represent curvature in the extreme, where the geometry folds inward on itself, but the same principle can stretch outward too. The rubber sheet only shows one side of that story.

The analogy also gives the impression that massive objects sit *on top* of space, deforming it from the outside. In truth, they're *within* space-time, and their mass and energy are part of what defines its structure. There isn't an empty sheet waiting to be distorted by things placed upon it. The fabric and the objects it contains are inseparable; they define each other. The presence of a star or planet doesn't just make a dent—it's part of the reason space-time exists in the form it does.

One of the most important takeaways from examining the rubber sheet analogy is how much it reveals about human understanding. Physics often starts with pictures like these, stepping stones that help people connect abstract math to intuition. Every analogy simplifies something complex to make it digestible. The trick is remembering that an analogy is never the thing itself. The rubber sheet helps us grasp how geometry replaces force in explaining gravity, but it must eventually give way to the deeper realization that the true shape of the universe can't be fully visualized in two or three dimensions.

Interestingly, when scientists teach relativity, they sometimes show how the analogy evolves. For basic understanding, the rubber sheet works. To explain black holes, they might use animations showing a funnel shape that extends infinitely deep, symbolizing how space curves near the event horizon. To describe the universe's expansion, they use the image of a balloon with dots representing galaxies—each dot moving away from the others as the surface stretches. These are all two-dimensional projections of a four-dimensional reality. They're useful tools, but each carries the same caveat: they're approximations, not literal depictions.

What makes black holes especially challenging to visualize is that their curvature is not just about bending—it's about stretching, twisting, and warping both space and time together. The rubber sheet shows one aspect of that bending, but not how the geometry changes the flow of time or how it can trap light itself. It's like trying to describe a song by showing someone a picture of the sound waves—it conveys structure but misses the experience.

7

THE EDGE OF UNDERSTANDING

At the heart of modern science are two towering achievements that have changed how we see the universe: Einstein's general relativity and the framework of quantum mechanics. Each describes a part of reality with astonishing precision, yet they don't agree with each other. It's as if nature has given us two perfectly crafted maps—one for the very large and one for the very small—but when we try to overlap them, the edges refuse to line up. Black holes sit right at the intersection of those maps, where the differences between them can't be ignored.

To understand why this clash exists, it helps to look at what each theory is trying to describe. General relativity is the language of the cosmos on grand scales. It explains how planets orbit stars, how galaxies form, and how gravity shapes space and time. It's smooth, continuous, and elegant—space and time behave like a flexible fabric that curves under the influence of mass and energy. Everything in it follows well-defined paths, determined by geometry.

Quantum mechanics, on the other hand, deals with the smallest building blocks of nature—particles, atoms, and the strange rules that govern their behavior. It's a world where nothing is smooth or certain. Instead of predictable paths, particles exist in clouds of probability. They can be in multiple states at once, leap between energy levels, and even influence one another instantly across vast distances. It's a framework of uncertainty, governed by chance and statistical outcomes rather than certainty and geometry.

Each theory has been tested to an extraordinary degree and has never failed within its own domain. Relativity correctly predicts how GPS satellites must adjust for time dilation to stay accurate, how light bends around stars, and how galaxies evolve. Quantum mechanics explains everything from the glow of a light bulb to the structure of DNA, the operation of computers, and the behavior of subatomic particles in giant colliders. Both are cornerstones of modern physics—and yet, when applied to the same situation, they refuse to coexist peacefully.

The trouble begins when gravity and quantum effects both become important, such as inside black holes or during the first moments after the Big Bang. Relativity describes gravity as a smooth curvature of space-time. Quantum mechanics insists that, at tiny scales, everything—including space and time—is grainy, fluctuating, and uncertain. When physicists try to merge the two descriptions, the equations break apart. Terms that should yield finite results instead explode into infinities. Instead of producing a coherent picture, the math collapses under its own weight.

One way to think about it is this: relativity treats the universe like a grand, continuous stage on which everything unfolds, while quantum mechanics treats the stage itself as part of the show—constantly shifting, vibrating, and uncertain. When you try to describe how the actors move across that unstable floor, the two scripts start contradicting each other.

For example, relativity tells us that space-time can curve smoothly around massive objects, while quantum theory says that, at the smallest scales, there's no such thing as a smooth surface. Space itself should be jittering with energy fluctuations—tiny tremors known as quantum foam. In that environment, relativity's smooth geometry can't exist. Likewise, relativity's idea of a singularity, a point of infinite density at the center of a black hole, makes no sense in quantum physics, where infinities are forbidden and probabilities rule. Each theory demands a universe that contradicts the other's assumptions.

Physicists have spent nearly a century trying to reconcile the two, and several approaches have emerged, each attempting to bridge the divide. One of the best-known is **string theory**, which replaces point-like particles with tiny, vibrating strings. In this view, gravity and quantum behavior can coexist because those strings stretch and oscillate, giving rise to both space-time and matter. The theory has mathematical beauty and internal consistency, but it's also extraordinarily complex—and, so far, unproven by experiment.

Another approach is **loop quantum gravity**, which starts from the opposite direction. Instead of trying to quantize particles, it tries to quantize space-time itself. It proposes that space is made of tiny, discrete loops woven together like a network. Gravity emerges from how those loops interact, making the fabric of space-time itself a kind of quantum system. This approach avoids infinities and preserves some of relativity's geometry, but it too remains theoretical, with little direct evidence.

There are also hybrid ideas—models that borrow parts

of both and attempt to apply them to specific problems, such as Hawking radiation or the early universe. Each of these theories is a step toward unification, but none has yet achieved what physicists dream of: a single framework that describes everything, from the motion of galaxies to the dance of subatomic particles.

The black hole remains the ultimate testing ground for this quest. At the event horizon, gravity is extreme, but quantum effects are still relevant. Near the singularity, both dominate completely. Any theory that hopes to unite physics must work there. When Stephen Hawking proposed that black holes emit radiation, it was one of the first hints that a bridge might exist. His work showed that quantum effects could interact with curved space-time, creating real, measurable outcomes. Yet, even that idea raised new questions—chief among them the information paradox, which revealed just how deep the clash runs.

The paradox goes like this: quantum mechanics says information about the state of particles can never be destroyed, while relativity seems to suggest that anything falling into a black hole is lost forever. When Hawking's theory predicted that black holes could eventually evaporate, that created an impossible situation. If they vanish, what happens to the information inside? If it's truly gone, quantum mechanics breaks. If it's somehow preserved, relativity's rules about horizons and space-time must be wrong. The two frameworks, once again, stand in direct opposition.

This conflict isn't just a matter of abstract equations—it touches on the deepest ideas about reality. Relativity sees the universe as deterministic: given enough information, you can predict everything that will happen. Quantum mechanics, by contrast, builds chance into the fabric of nature itself. Even if you know everything about a system, you can only predict probabilities, not certainties. These are fundamentally different ways of describing existence. One paints a picture of a universe that runs like clockwork; the other, of a universe that flickers with uncertainty at every level.

Yet both must be true, because both describe aspects of the same reality. That's what makes their tension so powerful. When two models describe the world perfectly but contradict each other, it means we're missing something deeper—a hidden principle that contains both. History shows that every time science hits such a contradiction, a revolution follows. Newtonian mechanics gave way to relativity when light refused to obey its rules. Classical physics gave way to quantum mechanics when atoms refused to behave predictably. The clash between relativity and quantum theory may well be the next great turning point.

Some physicists think the answer lies in finding a more complete version of quantum theory—one that naturally includes gravity. Others think we need to rethink space and time altogether, that they might be emergent properties rather than fundamental ones. There are even ideas suggesting that the universe itself could be a kind of hologram, where the information describing everything that happens in three dimensions is encoded on a two-dimensional surface. Each of these possibilities points toward the same goal: unification.

It's easy to think of this as a battle between theories, but in truth, it's more of a dialogue. Relativity and quantum mechanics aren't enemies—they're perspectives that describe different layers of the same reality. When scientists talk about the "great clash," they're really describing a growing tension in understanding that forces progress. Every time we test relativity with gravitational waves or explore quantum effects in new materials, we're gathering data that may one day reveal how the two connect.

At its core, the conflict between these theories is a reminder that the universe isn't simple, and that our understanding of it is still evolving. Physics doesn't aim to provide absolute answers—it aims to build better questions. Black holes, with their merging of extremes, keep those questions alive. Somewhere in the way they warp space and tangle time lies the bridge between the smooth and the chaotic, between certainty and probability.

That bridge, once found, won't just change how we think about black holes. It will redefine how we think about everything—space, time, energy, even reality itself. For now, we stand between two brilliant but incomplete maps, tracing the faint outlines where they might one day meet. The clash between relativity and quantum mechanics isn't a deadlock; it's the sound of physics stretching toward something larger, something that may finally reveal the true architecture of the universe.

THE INFORMATION PARADOX: DOES INFORMATION DISAPPEAR?

Chapter 9 – The Information Paradox: Does Information Disappear?

Among all the puzzles that black holes have presented to physicists, none has been as haunting or as provocative as the information paradox. It's not just a question about astronomy or equations—it's a question that strikes at the foundations of what it means for the universe to make sense. It asks something simple on the surface: when something falls into a black hole, is the information about what it was—its structure, state, and composition—gone forever? The trouble is, both of our best physical theories give opposite answers, and neither can be dismissed.

In physics, "information" doesn't mean gossip or trivia—it means the detailed description of a system's state. Every atom, every particle, every bit of energy has properties that define it: its position, momentum, spin, and so on. Together, those properties make up the information that describes the system. According to the rules of quantum mechanics, that information can never truly vanish. It might change form, get scrambled, or become impossible to reconstruct, but it can't be destroyed. That's a fundamental principle.

Now consider what happens in relativity. When matter falls into a black hole, the event horizon becomes the dividing line between what's observable and what isn't. Anything crossing that boundary is cut off from the rest of the universe. The information describing it can no longer influence anything outside the black hole. To an external observer, it's effectively gone. If the black hole later evaporates through Hawking radiation, the question becomes: where did the information go? The radiation appears random, carrying no details about what originally fell in. And that's where the paradox begins.

Stephen Hawking's calculations in the 1970s showed that black holes aren't completely black. Quantum effects near the event horizon cause them to emit faint radiation, slowly losing mass over time. This process—Hawking radiation—suggested that black holes could eventually evaporate entirely. But if they disappear, what happens to all the information they once contained? If it's truly lost, that would violate one of quantum theory's most sacred rules. If it's somehow preserved, then our understanding of spacetime and horizons must be missing something profound.

Either way, one of our cornerstone theories must be incomplete.

This paradox isn't just an academic debate. It forces us to confront how we define reality. In classical physics, the universe behaves like a perfectly written record book—every event leaves a trace, and those traces, in principle, could always be read. In quantum mechanics, that principle survives even when randomness enters the picture. You might not be able to predict exactly what will happen, but you can always, in theory, work backward and figure out what did. The universe keeps its books balanced. The information paradox threatens to rip out a few pages and burn them.

Physicists immediately began searching for ways to resolve the problem without breaking either theory. One idea was that the information isn't destroyed—it's just trapped inside the black hole, inaccessible to the outside universe. That would save quantum mechanics but leaves another issue: if the black hole evaporates completely, the trap disappears. The information has nowhere to go. That means something else must happen before the end—something that lets the information escape in a way we don't yet understand.

Others proposed that the information might be encoded somehow on the event horizon itself, rather than inside the black hole. This idea led to what's known as the **holographic principle**, one of the most important concepts in modern theoretical physics. It suggests that the information about everything that falls into a black hole could be stored on its surface, just as a hologram on a flat sheet of film can represent a three-dimensional image. The interior, in a sense, might be an illusion—a projection of data encoded at the boundary.

If that's true, it changes how we think about reality at a fundamental level. It means that all the information describing a region of space could be represented on a lower-dimensional surface. The universe itself, in some interpretations, could function like a vast hologram. The concept is elegant and consistent with quantum theory, but it also raises more questions than it answers. How does that surface store the information? How is it released when the black hole evaporates? Can it be recovered in any meaningful way, or is it just a bookkeeping trick to preserve the math?

Another line of reasoning came from looking more closely at Hawking radiation. Some physicists proposed that perhaps the radiation isn't truly random—that subtle quantum correlations between the emitted particles might encode the missing information in ways that are incredibly hard to detect but still present. If that's true, the black hole doesn't destroy information; it simply releases it in a form so scrambled that it looks random to us. In that case, the universe would preserve its ledger after all, just written in an indecipherable code.

However, this idea also comes with complications. If information escapes in radiation, it means something near the event horizon must allow those quantum states to influence the outgoing particles. But relativity insists that nothing can cross that boundary outward. That led to yet another concept—one that's far more controversial—called the **firewall hypothesis.** It suggests that the event horizon might not be a smooth, invisible boundary after all. Instead, there could be a violent, energetic barrier that burns up anything falling in, releasing its information outward. But this contradicts the smooth geometry predicted by general relativity and undermines the very definition of a black hole. It trades one paradox for another.

These competing ideas highlight the depth of the conflict between quantum mechanics and relativity. Neither theory can comfortably accommodate the other's demands without giving something up. Relativity wants smooth, continuous space-time with no abrupt boundaries. Quantum theory demands that information be preserved at all costs. The information paradox forces physicists to choose which principle to bend. Yet every attempt to bend either one leads to deeper inconsistencies.

Despite decades of debate, no consensus has emerged. The holographic principle remains a strong candidate, supported by mathematical work in areas like string theory, where the idea of information living on boundaries fits neatly into the equations. But other physicists remain skeptical, arguing that while it preserves information in theory, it doesn't explain how that information becomes accessible again in a physical sense. For them, the mystery remains unresolved.

What's striking is how this single paradox has reshaped modern physics. It's not just about black holes anymore—it's about the nature of reality, causality, and even time itself. If information truly can disappear, then the universe might be fundamentally unpredictable in ways we can't yet grasp. If it can't, then our picture of space and gravity must change to accommodate that truth. Either outcome would revolutionize science.

Some physicists think the answer might come from quantum gravity—the long-sought theory that merges the two great frameworks into one consistent picture. Such a theory would describe both the geometry of space-time and the quantum behavior of matter in a unified way. It might show that the paradox isn't really a contradiction but a misunderstanding of how these systems work together. For now, though, that theory remains out of reach.

The paradox persists not because we lack intelligence or data, but because it touches something deep and unresolved in the structure of our theories. It's a reminder that every great discovery opens new layers of mystery. Hawking himself changed his stance late in life, suggesting that perhaps information could leak out after all—but even then, the mechanism remained uncertain.

STEPHEN HAWKING'S INSIGHTS

In the early 1970s, physics found itself at an unusual crossroads. Einstein's theory of general relativity had reshaped how scientists thought about gravity and the large-scale structure of the universe. Quantum mechanics, meanwhile, had revealed the strange and unpredictable world of particles and energy at the smallest scales. Both theories were brilliant—but incompatible. Then came a young physicist named Stephen Hawking, who would take that clash and use it to uncover one of the most surprising truths about black holes: that they aren't entirely black.

At the time, black holes were thought to be one-way traps. They could swallow anything—matter, light, radiation—but nothing could ever come out. They were the ultimate sinkholes of the cosmos, where information went to die. That view fit perfectly with Einstein's equations, which described black holes as pure geometry: regions where gravity was so intense that nothing, once past the event horizon, could escape. But Hawking began to wonder what would happen if you looked at black holes not just through

Einstein's eyes, but through the lens of quantum theory as well.

Quantum mechanics has a very different view of empty space. What looks like nothing isn't truly empty—it's a restless sea of energy. Even in a perfect vacuum, tiny pairs of particles and antiparticles constantly appear and disappear. These are called quantum fluctuations. They exist for a fleeting moment before annihilating each other and returning their borrowed energy back to the universe. Under normal conditions, this dance of creation and destruction happens everywhere, unnoticed. But near the edge of a black hole, that balance changes dramatically.

Hawking realized that if such a particle pair formed right on the edge of a black hole's event horizon, strange things could happen. Normally, the two particles would collide and cancel each other out almost immediately. But what if one fell into the black hole while the other escaped before they could reunite? The one that escaped would fly off into space as real radiation, while the other, now trapped inside the black hole, would effectively carry negative energy from the perspective of the outside universe.

This process meant that the black hole would lose a tiny bit of its mass over time. The energy carried away by the escaping particles would come at the expense of the black hole's own energy. In other words, the black hole would slowly evaporate. This idea was revolutionary. Black holes, once thought to be eternal, could actually shrink and disappear—radiating heat as they did. The radiation predicted by this process became known as **Hawking radiation.**

The implications of this discovery were enormous. It was the first time quantum mechanics and general relativity had been successfully combined to describe the same physical process. Hawking's insight showed that the edge of a black hole wasn't just a silent boundary—it was an arena where quantum and gravitational effects intertwined in ways that challenged both theories. It also transformed black holes from being pure absorbers of energy to participants in a cosmic exchange, slowly releasing their mass back into the universe.

When Hawking first presented his results, even other physicists were skeptical. The idea that something could escape a black hole seemed to contradict everything they knew. But as the math was checked and rechecked, it became clear that Hawking's logic held. He hadn't found a loophole in the laws of physics—he'd found a bridge between them. His calculations showed that black holes have temperature and entropy, two concepts usually reserved for ordinary physical systems. That meant they weren't just abstract geometric objects but thermodynamic ones, with measurable properties linked to heat and energy.

The idea of black hole temperature may sound abstract, but it carries deep meaning. Temperature, in physics, is a measure of how much random motion or disorder exists within a system. Entropy, a related concept, describes the number of possible ways that system's particles can be arranged while still appearing the same overall. When Hawking showed that black holes have entropy, he implied that they have internal states—a hidden complexity beneath their smooth exterior. That was startling, because it meant black holes weren't featureless voids after all. Something inside them held the key to how information was stored, processed, or perhaps even released.

This connection between entropy and the area of a black hole's event horizon became one of the most elegant results in physics. It suggested that the surface area—not the volume—was what determined how much information a black hole could contain. This simple but profound relationship would later inspire the holographic principle, the idea that the information inside a region of space can be described by data on its surface. In a sense, Hawking's work didn't just change how we think about black holes—it changed how we think about space itself.

Still, Hawking radiation came with an unsettling twist. If black holes can evaporate, what happens to the matter—and the information—they've consumed? When a black hole radiates away its mass, the radiation seems random, containing no hint of the information about what fell in. That led directly to the information paradox, the puzzle of whether the universe's most fundamental rule—that information cannot be destroyed—might actually be violated. Hawking himself initially believed it was. In his early papers, he argued that information was lost forever when a black hole evaporated. It wasn't a comforting thought, but it was consistent with his calculations.

Over time, however, that view began to shift. Other physicists, including those working in quantum field theory and string theory, started exploring whether subtle correlations might exist in the radiation—patterns too faint to notice but still preserving the information. Hawking, after decades of debate, eventually conceded that they might be right. Information, he suggested later in life, could escape after all, though the mechanism remained a mystery. That humility—to change his stance based on new reasoning—became as much a part of his legacy as the discovery itself.

What makes Hawking's insight so powerful isn't just that it solves a particular problem, but that it reframes how we think about extremes. Before his work, black holes represented dead ends for science—places where equations failed and understanding stopped. After Hawking radiation, they became laboratories for testing the deepest questions in physics. They linked gravity, quantum mechanics, and thermodynamics into a single framework. They also gave us a rare glimpse of unity between theories that otherwise seem to live in separate worlds.

The elegance of Hawking's idea lies in its simplicity. He didn't need new equations or exotic assumptions—just the courage to apply existing principles in an unfamiliar place. By asking what quantum mechanics would look like on the curved background of space-time near a black hole, he found something no one had dared to expect: that the emptiest places in the universe could glow with energy. That glow might be too faint to detect for most black holes—far weaker than the cosmic microwave background that fills space—but it's there in principle, written into the fabric of the universe itself.

In a broader sense, Hawking radiation also changed how scientists think about boundaries and permanence. It showed that even the most absolute-seeming limits—like the event horizon—can blur under closer inspection. Nature, it turns out, doesn't deal in absolutes as often as we think. Where we see an unbreakable wall, physics often finds a process that bends the rules just enough to make them richer.

Hawking's discovery continues to influence nearly every field of theoretical physics today. It underpins the study of quantum gravity, shapes our understanding of cosmic entropy, and challenges our assumptions about time and causality. Each generation of scientists builds on his insight, refining and reinterpreting it through new frameworks. Whether it's in the context of string theory, loop quantum gravity, or holography, Hawking's idea remains a touchstone—a reminder that the universe still has secrets that can be revealed through thought as much as through experiment.

WHY BLACK HOLES MIGHT NOT LAST FOREVER

For most of history, black holes were thought to be the ultimate cosmic endpoints—objects that swallow everything and never let anything escape. They were treated as eternal prisons of gravity, existing for all time, unchanging once formed. But Stephen Hawking's discovery that black holes can emit radiation transformed that idea. It introduced a strange and unsettling possibility: that even these giants might not be forever.

To understand why, it helps to start with what makes black holes so stable in the first place. They're born from collapse—when a massive star runs out of fuel, its core contracts under its own gravity. If it's heavy enough, the collapse becomes unstoppable, crushing matter to a point of infinite density, wrapped in an event horizon that nothing can cross. Once formed, there's nothing in the known universe that can push back strongly enough to undo that. A black hole just sits there, pulling in whatever strays too close. It seems indestructible.

But physics doesn't like true permanence. Every known process, from the decay of atoms to the expansion of galaxies, involves change. Even in the deepest vacuum of space, things fluctuate and move. Hawking's insight that black holes radiate heat was the first hint that they might also evolve—slowly, quietly, and over unimaginable timescales. The key lies in the balance between energy and gravity.

According to his calculations, black holes emit radiation because of quantum effects near the event horizon. The details are subtle, but the outcome is straightforward: energy leaks away. That energy must come from somewhere, and the only source available is the black hole's own mass. Over time, that means the black hole loses weight—bit by bit, particle by particle.

For the kinds of black holes that form from stars, this process is glacially slow. A black hole with the mass of our Sun would take roughly 10^{67} years to evaporate completely. That's a number so vast it's nearly meaningless—far longer than the current age of the universe, which is around 10^{10} years. But given enough time, even something that seems eternal can fade.

As a black hole loses mass, it grows smaller. And as it grows smaller, it gets hotter. That's the opposite of how most objects behave. A cooling cup of coffee loses energy and becomes colder. A shrinking black hole does the reverse—it radiates faster as it contracts. The smaller it becomes, the more intense its radiation gets, speeding up the process in a kind of feedback loop. In its final moments, it wouldn't fade quietly—it would end in a brief but spectacular burst of energy, releasing the last of its mass in a flash.

The idea that black holes could evaporate changes how we think about the distant future of the universe. If nothing can last forever, what does that mean for the cosmos itself? Over trillions upon trillions of years, stars will burn out, galaxies will drift apart, and only remnants—white dwarfs, neutron stars, and black holes—will remain. If Hawking's prediction holds, even those remnants will eventually vanish, leaving behind only radiation and sparse particles drifting through an ever-colder, ever-emptier space.

That far-off future, sometimes called the "black hole era," represents the slow winding-down of everything. The last lights of the universe won't be stars but evaporating black holes, flickering one by one into darkness. It's a haunting but strangely beautiful vision—a cosmos that ends not in violence, but in quiet dissolution.

Of course, that future is theoretical. Hawking radiation has never been directly observed. For stellar or supermassive black holes, it's far too faint to detect. Their radiation is completely drowned out by the cosmic background glow of the universe. Only hypothetical tiny black holes—ones that might have formed shortly after the Big Bang—would evaporate quickly enough to reveal the effect directly. None have been found, but the search continues.

Still, even without direct evidence, the concept is deeply rooted in both quantum mechanics and thermodynamics. In many ways, it feels inevitable. If black holes have temperature, and anything with temperature radiates energy, then they must lose mass. Nature doesn't make exceptions to that rule. The real question isn't whether they can evaporate, but how—and what happens to the information trapped inside them when they do.

The fate of that information remains one of physics' deepest riddles. If the black hole truly disappears, taking the data about what fell into it along with it, that would violate one of quantum mechanics' fundamental laws. But if the information somehow escapes, it challenges our understanding of space-time and causality. Either outcome forces a change in how we think about reality.

Some scientists have proposed that as black holes evaporate, the information they hold might be released in the final stages, encoded in the radiation itself. Others suggest that the information remains imprinted on the event horizon, gradually leaking out as the black hole shrinks. There's even speculation that when a black hole reaches the smallest possible scale, it might leave behind a stable remnant—a tiny, dense object holding the information forever. Each idea offers a different way for the universe to keep its accounting balanced.

Beyond the physics, there's something poetic about the idea of black holes dying. They begin as destroyers—objects that devour everything in reach. Yet, in the end, they give back their energy to the cosmos. What they took, they eventually return, scattered across the universe as faint radiation. Their death isn't an act of destruction, but a kind of balance being restored.

The timescales involved are almost beyond comprehension. To put it into perspective, if the entire age of the universe so far were compressed into a single second, the full evaporation of a black hole would take longer than billions of entire universes. It's an event so far in the future that it defies human intuition. But physics doesn't care about human timescales—it cares about consistency. Given infinite time, every process that can happen, does.

This understanding also reshapes how cosmologists think about permanence. Black holes once symbolized the end of matter, the ultimate sink from which nothing returns. Hawking's insight turned that narrative on its head. The universe, it turns out, has no true dead ends—only cycles that play out on scales of time and energy far beyond what we can witness.

For now, black holes appear eternal to us. Even the smallest ones we've detected will outlive every star that exists today. The massive ones at galactic centers—millions or billions of times the Sun's mass—will persist even longer. But Hawking's theory tells us that change, however slow, is still inevitable. The same physics that governs the decay of atoms applies, in principle, to the collapse of galaxies.

When that process is carried to its ultimate conclusion, we're left with a universe stripped of its structure—no stars, no planets, no light. Just a thin bath of low-energy radiation drifting through an endless void. In that picture, even black holes aren't final—they're transitional states in the universe's long journey toward equilibrium.

Hawking's prediction didn't just reveal that black holes might die; it redefined what it means for something to exist forever. It showed that eternity in physics is an illusion—that everything, no matter how massive or seemingly unchanging, participates in the slow unfolding of time and energy.

WHAT THEIR EVENTUAL EVAPORATION WOULD MEAN

If black holes can evaporate, that single fact forces a rethinking of nearly everything we assume about permanence, time, and the universe's fate. They've always seemed like the ultimate symbols of endurance—unchanging monuments of collapsed stars and cosmic violence. Yet the possibility that even they can fade tells us something humbling: the universe doesn't do "forever." It does transformation.

When we talk about a black hole's eventual evaporation, we're talking about an event that unfolds across unfathomable stretches of time. For a massive black hole, the process would take longer than the age of the universe multiplied by trillions upon trillions. Nothing we know could witness it. But the meaning of that future isn't just about time—it's about what it says regarding the destiny of everything.

In the grand story of the cosmos, black holes are often considered the last players left on stage after all others have exited. Stars will burn through their fuel and die. Galaxies will spread farther apart as space continues to expand. Even the faint glow of background radiation will cool to near nothing. What remains are black holes—quiet, invisible, and scattered across a vast, dark universe. But if Hawking radiation is correct, even they won't last.

As black holes evaporate, they release energy back into space. It's faint and slow at first, but that radiation represents a return—matter that had once been locked away is now being redistributed. That process means black holes are not simply destroyers, but recyclers. They take in the universe's excess and eventually give it back in a new form. Over countless eons, this release would change the composition of the cosmos itself. What was once dense and hot will become thin and cold. Energy that once formed stars and galaxies will spread out into near-nothingness, a final settling of the universe's accounts.

The meaning of that shift goes beyond physics. It suggests that even the most extreme forms of matter and gravity aren't immune to time's erosion. The great engines that once shaped galaxies will themselves yield to entropy—the tendency for energy to spread and even out. That process, called heat death, describes a universe that continues to expand and cool until no usable energy remains. Black hole evaporation isn't a side note to that story—it's the final act.

When the last black hole releases its final burst of radiation, there will be nothing left to generate new structure. No stars. No collisions. No new heat. The cosmos will be a smooth sea of faint particles, expanding endlessly into silence. It's an eerie picture, but it also reflects a kind of cosmic symmetry. Everything that once was—matter, light, even the energy that powered gravity itself—will have completed a cycle from order to disorder, from brightness to stillness.

Some physicists see this not as an ending but as balance. The energy doesn't disappear; it only spreads. Matter is converted into radiation, and that radiation disperses evenly across an infinite horizon. The laws of conservation are upheld. Nothing is truly gone, just diffused beyond recognition. In that sense, evaporation is a return to equilibrium—the universe reaching a state where everything is as evenly distributed as it can possibly be.

There's also a more subtle meaning in this process: the story of information. Every object that falls into a black hole carries information about its composition and history. If black holes truly evaporate, what becomes of that information? Does it disappear into nothing, or is it somehow encoded in the radiation that escapes? Physicists are still wrestling with that question, because its answer touches on the very foundation of what reality is. If information is preserved, then even the last light from the last black hole contains the memory of everything that came before. If it's lost, then the universe forgets.

Either possibility has weight. A universe that remembers is one where nothing is ever truly erased—only transformed beyond our ability to recognize it. A universe that forgets, on the other hand, is one where meaning itself may be temporary. Hawking's later work suggested that perhaps the information does survive, though scrambled in ways we can't yet decode. That idea—that nothing is truly lost—adds a quiet note of continuity even at the end of all things.

The evaporation of black holes also reframes how scientists think about the nature of space and time. If even the most stable structures dissolve, then permanence may not exist anywhere in the cosmos. Every object, every pattern, even the fabric of space-time, might be part of a long sequence of change. Nothing is static; everything is part of a vast process of unfolding.

From a human perspective, it's hard to feel connected to events that happen over timescales so large they make geological ages look like flickers. Yet the fact that physics can describe them at all is remarkable. It means the same laws that govern the fall of an apple or the glow of a lightbulb can also tell us something about the death of black holes trillions of years from now. That unity across scale is one of the quiet triumphs of science.

If black holes truly vanish, what happens to their gravity? When a black hole loses mass, its gravitational pull weakens proportionally. As it evaporates, the space around it flattens out. The distortion in space-time that defined it fades until, eventually, it's as if it was never there. That disappearance would leave no trace—no crater, no shell, no remnant. The warping of space itself would relax back into smoothness, restoring the fabric it once bent.

It's a strangely poetic image: the universe slowly ironing out its wrinkles. Every collapsed star, every distortion, every gravitational well, undone by the quiet passage of time. In the end, what remains isn't chaos but calm. Space-time becomes flat and featureless, an infinite blank canvas waiting for nothing.

There's another way to look at it, though—not as an ending, but as the resetting of conditions. Some theories suggest that quantum fluctuations—the same kind that gave rise to the earliest structures after the Big Bang—could one day spark new regions of expansion, perhaps even new universes. If that's true, the end of one cosmic cycle might carry the seeds of another. The evaporation of black holes could be part of a rhythm much larger than we can see.

8

THE GIANTS IN THE GALACTIC CENTER

At the very center of the Milky Way—beneath the swirl of hundreds of billions of stars, through clouds of gas and dust that stretch for light-years—there's an object that ties it all together. It's invisible to the eye, but its presence defines the structure and motion of our entire galaxy. This is Sagittarius A*, the supermassive black hole at the Milky Way's heart.

It's not the largest black hole in the universe, nor the most active. Compared to the brilliant quasars found in distant galaxies, it's quiet—almost sleepy. Yet it holds extraordinary significance, both scientifically and symbolically. It's the closest supermassive black hole we can study, and through it, we've been able to confirm some of the most astonishing predictions of modern physics.

For much of human history, the center of our galaxy was nothing but a patch of stars in the direction of the constellation Sagittarius. Early astronomers saw it as a thick, glowing band cutting across the sky, not realizing that this was the dense plane of our own galaxy seen from within. The idea that something massive and invisible lay hidden there didn't emerge until the twentieth century.

By the 1970s, radio astronomers began detecting an unusually strong signal coming from that region. The source appeared compact—smaller than our solar system—but emitted radio waves that hinted at something powerful and mysterious. They named it Sagittarius A*, with the "A" marking it as the brightest source in that part of the sky and the "*" indicating it was a distinct, variable object.

At first, astronomers weren't sure what they were looking at. Some proposed that it might be a dense cluster of neutron stars or a collection of stellar remnants tightly packed together. But as telescopes improved, especially in the infrared spectrum, researchers began tracking the movement of individual stars near the galactic center. What they found left no room for doubt.

These stars weren't drifting lazily. They were racing around something unseen at incredible speeds—thousands of kilometers per second. One star, now known as S2, moves in a 16-year orbit that brings it perilously close to the center. By mapping its motion precisely, scientists were able to calculate the mass of whatever it circled. The result was staggering: about four million times the mass of the Sun, all concentrated in a region smaller than Mercury's orbit. Nothing but a black hole could explain that.

Sagittarius A* doesn't draw attention the way some galactic centers do. It's not engulfing massive streams of gas or lighting up with energy jets. In cosmic terms, it's relatively quiet. But every once in a while, it flares briefly—perhaps when a cloud of gas or a small asteroid strays too close and is shredded by tidal forces. The debris heats up as it spirals inward, releasing X-rays before vanishing beyond the event horizon. Each of these flickers is a reminder that, even dormant, Sagittarius A* is far from inactive.

The ability to observe it directly was a long-standing dream for astronomers. For decades, no telescope was powerful enough to resolve details that small and that far away. The region is buried behind dense interstellar dust that blocks visible light. But in 2022, the Event Horizon Telescope—a global network of synchronized radio observatories—finally succeeded in producing the first image of Sagittarius A*. What it revealed was not the black hole itself, but a bright ring of radiation surrounding a dark central shadow. That ring marks the photon orbit, the last stable path light can take before being swallowed.

The image confirmed that the object behaves exactly as Einstein's equations predicted. Space-time around it is curved in the precise way general relativity describes. Even though black holes were once theoretical constructs, the picture of Sagittarius A* proved they're as real and measurable as any other part of the universe.

What's perhaps most striking is how this immense object coexists so peacefully with its surroundings. The stars near it aren't constantly being devoured; they orbit safely at distances that keep them stable for millions of years. Our solar system, about 26,000 light-years away, feels its gravitational influence only indirectly—just enough to help define the galaxy's structure. If the black hole vanished overnight (which it won't), the Milky Way would remain largely unchanged. Its mass, though vast, is only a small fraction of the galaxy's total.

Still, Sagittarius A* acts as the gravitational anchor of our spiral galaxy. It gives the Milky Way its balance and shape. The stars, nebulae, and dust that orbit the center all respond to its pull, maintaining the slow cosmic dance that defines our galactic home. It's a stabilizer on an unimaginable scale.

Studying Sagittarius A* has given astronomers more than just a deeper understanding of black holes—it's also offered a way to test the boundaries of physics. When stars like S2 swing close to the black hole, their light is redshifted by the immense gravity, a measurable stretching of wavelengths caused by time itself running slower near the event horizon. Observing these effects allows scientists to verify how space and time behave under extreme conditions—something no laboratory on Earth could reproduce.

Even more revealing are the orbits of smaller stars and gas clouds that wander through the central region. Their movements act like tracers, mapping the invisible landscape of gravity. Every new observation tightens our understanding of Sagittarius A*'s size, spin, and environment. Researchers now think it might rotate, twisting space-time around it in a phenomenon called frame-dragging. If that's true, it would subtly affect how nearby matter moves, an effect scientists are still working to measure precisely.

Despite being the closest supermassive black hole, Sagittarius A* remains elusive in some ways. It's hard to predict when it will flare, and its dimness compared to brighter galactic centers makes it a challenge to study in detail. Yet that quiet nature is also a gift—it allows scientists to observe the dynamics around it without the blinding interference of intense radiation.

Every insight drawn from Sagittarius A* ripples outward into broader questions. How do black holes and galaxies grow together? What determines whether a black hole is calm or active? And how does matter behave at the edge of space-time? The Milky Way's central black hole provides the perfect natural laboratory for exploring those mysteries.

HOW SUPERMASSIVE BLACK HOLES SHAPE GALAXIES

At first glance, it might seem strange that something as destructive as a black hole could play a creative role in the universe. Yet, that's exactly what supermassive black holes do. These colossal objects, hidden deep in the centers of galaxies, help sculpt the very structures we see through telescopes. Their gravity, energy, and influence extend far beyond the region where light vanishes. Without them, galaxies might not look—or even exist—the way they do.

The link between a galaxy and its central black hole is not a casual one. When astronomers began measuring galaxies carefully, they found a surprising relationship: the bigger a galaxy's central bulge of stars, the more massive its black hole tends to be. That connection suggests the two grow together, influencing each other across cosmic time. It's as if the galaxy and its black hole are partners in an elaborate dance, each shaping the other's steps.

In the early universe, when galaxies were young and chaotic, gas clouds were everywhere—dense, hot, and ready to collapse into stars. Some of that gas inevitably found its way toward the galactic center, where gravity pulled it inward. As it spiraled toward the forming black hole, it heated up, releasing enormous energy. The radiation and jets from this process blasted outward, pushing on the surrounding gas. Instead of simply swallowing everything, the black hole's activity sent ripples through its entire galaxy.

Those ripples weren't gentle. The energy streaming from an actively feeding black hole can be intense enough to heat or expel gas from the galaxy's inner regions. That might sound like a destructive act, but it's actually a form of regulation. If too much gas were to collapse at once, the galaxy would burn through its fuel too quickly, forming too many stars too fast. By pushing some of that gas away, the black hole slows down the process, ensuring the galaxy develops more evenly over time. It's a cosmic balancing act—gravity draws matter in, while radiation and jets push some of it back out.

The most dramatic examples of this interaction are seen in quasars and active galactic nuclei. These are galaxies whose central black holes are devouring material at incredible rates. The light and radiation from these regions can outshine all the stars in the galaxy combined. While this energy can blast gas outward for thousands of light-years, it also stirs and compresses clouds farther away, sometimes triggering new bursts of star formation. Depending on how the energy interacts with the surrounding material, the black hole can either suppress or stimulate the birth of new stars.

Over millions of years, this back-and-forth shapes the galaxy's structure. In some cases, powerful jets from the black hole carve tunnels through interstellar gas, leaving behind vast cavities. These jets, often made of charged particles moving near the speed of light, stretch far beyond the visible galaxy. Observations have shown them extending for hundreds of thousands of light-years, forming enormous lobes of radio emission. They carry energy deep into intergalactic space, influencing not only their own galaxy but the surrounding cosmic environment as well.

One of the more surprising lessons from recent studies is that these feedback processes aren't rare—they're fundamental. Every large galaxy seems to have experienced periods when its central black hole was active. Some galaxies show evidence of past outbursts in the form of giant bubbles or shock fronts still expanding through their gas halos. Even our Milky Way bears traces of this history. Two massive structures known as the Fermi Bubbles, extending above and below the galactic plane, are thought to be remnants of ancient activity from Sagittarius A*, when it may have briefly flared into a quasar-like state.

Supermassive black holes also affect how galaxies evolve after their most active phases. Once the black hole has consumed or expelled much of the nearby gas, the galaxy often transitions into a quieter, more stable period. Star formation slows, and the galaxy's core becomes less luminous. Astronomers believe this process explains why large elliptical galaxies, which host the biggest black holes, tend to have older, redder stars and little new star formation. The black hole's feedback effectively turned off the galaxy's ability to make new stars long ago.

Yet, not all galaxies follow the same path. Spiral galaxies like the Milky Way still have reservoirs of gas that sustain steady star formation. Their central black holes, while massive, remain relatively calm. This variety shows that black holes aren't simply engines of destruction—they're regulators of galactic life. When they're active, they reshape their surroundings through radiation, magnetic fields, and winds. When they're quiet, they serve as gravitational anchors, holding the galactic structure together.

It's not just the large-scale structure that's affected. Even the motions of individual stars are influenced by the black hole's presence. In the central regions of galaxies, stellar orbits become more chaotic, forming dense bulges that contrast with the orderly rotation of the outer spiral arms. This difference arises from the complex gravitational field near the galactic core, where the black hole's mass dominates. Farther out, the galaxy's overall mass smooths things out, creating the graceful patterns we see in spiral galaxies.

Computer simulations have helped bring these ideas to life. By modeling the interactions between gas, stars, and black holes, scientists can watch virtual galaxies evolve over billions of years. These simulations show that without the energy released by supermassive black holes, galaxies would grow too large and dense, forming unrealistic numbers of stars. The feedback from black holes, in contrast, produces galaxies that match what we actually observe—those with balanced distributions of stars, realistic sizes, and the right mix of activity and calm.

Beyond shaping their own galaxies, supermassive black holes influence the space between them. In galaxy clusters, for example, the combined activity of central black holes helps regulate the temperature of the surrounding gas. Without that input of energy, the gas would cool rapidly and collapse, triggering runaway star formation across the cluster. Instead, the heat from these colossal engines keeps the intergalactic medium in a delicate equilibrium.

The idea that something invisible could sculpt structures on such vast scales is both astonishing and elegant. It transforms black holes from mere cosmic oddities into central players in the story of the universe. They connect the smallest scales—where quantum physics governs the behavior of particles near the event horizon—to the largest scales, where galaxies and clusters evolve over billions of years.

THE SURPRISING WAYS THEY MAY *CREATE* STARS, NOT JUST DESTROY THEM

The word "black hole" tends to conjure images of destruction—of light being swallowed, of matter torn apart, of cosmic doom. For a long time, that was the only story scientists told about them. They were the universe's ultimate takers, endlessly consuming and never giving anything back. Yet, as our understanding deepened, a more nuanced picture began to appear. Black holes, especially the supermassive ones at the centers of galaxies, aren't just destroyers. Under the right conditions, they can also help create.

It sounds contradictory at first. How could something defined by its ability to consume everything around it possibly lead to creation? The answer lies in the complex interplay between gravity, heat, and motion. When matter falls toward a black hole, it doesn't disappear instantly. Before crossing the event horizon, it gathers into an accretion disk—a spinning, chaotic whirlpool of gas, dust, and magnetic fields. That disk can grow incredibly hot, releasing radiation and driving powerful outflows of energy and particles. Those outflows, rather than being purely destructive, can compress surrounding gas clouds in ways that spark the birth of new stars.

It's a little like a wildfire in nature. On the surface, it seems devastating, consuming everything in its path. But after the blaze passes, the landscape it leaves behind can be richer, fertilized with ash and opened to sunlight. In the cosmic sense, the "fires" around black holes can set off cycles of renewal across vast distances.

One of the key mechanisms behind this creative process is something called **feedback**—the exchange of energy between the black hole and its galaxy. When a black hole feeds, it releases streams of radiation and matter at near-light speeds. These jets and winds blast outward, colliding with clouds of gas scattered throughout the galaxy. The impact can strip some gas away, slowing star formation in certain areas. But it can also have the opposite effect. When the shockwaves pass through cold, dense regions, they squeeze them just enough to make the gas collapse under its own gravity. That collapse is exactly how new stars are born.

In that sense, black holes can act as cosmic gardeners. Their violent outbursts prune and shape their environments, clearing away thin, unstable gas while compacting the regions that are ready to form stars. It's not a gentle process, but it's efficient. By regulating the flow of energy through a galaxy, a supermassive black hole can help control when and where stars appear.

Observations have confirmed that this dual role isn't rare. In some galaxies, especially those where the central black hole is active, astronomers have found streams of material shooting out from the core and colliding with nearby gas clouds. These collisions often lead to bright knots of newborn stars forming along the edges of the outflow. Instead of erasing stellar nurseries, the black hole's energy becomes the trigger that lights them up.

This process can unfold on truly enormous scales. In some galaxies, jets extend for hundreds of thousands of light-years, sculpting the interstellar medium far beyond the galaxy's visible edge. The tips of these jets often contain regions of concentrated star formation, like sparks flaring at the end of a cosmic torch. It's as though the galaxy is being reshaped by invisible hands, its raw materials rearranged into new patterns by the very object that might once have seemed intent on devouring them.

The balance between destruction and creation depends on subtle conditions—how much gas is available, how fast the black hole is feeding, and the density of the surrounding environment. When the outflows are too strong, they can blow gas completely out of the galaxy, halting star formation altogether. But when they're moderate, they act more like a sculptor's chisel, refining rather than erasing. It's this middle ground where creation thrives.

What's remarkable is how universal this relationship appears to be. In small galaxies and large ones alike, the fingerprints of black hole feedback are visible. Even in our own Milky Way, astronomers have detected signs that Sagittarius A* may have played a role in shaping the central regions of the galaxy. The vast structures known as the Fermi Bubbles—two enormous plumes of energetic gas extending above and below the galactic plane—hint at a past outburst that could have both cleared and compressed the gas near the core. While Sagittarius A* is quiet now, those echoes suggest it may once have helped spark waves of star formation in the galaxy's inner regions.

This idea—that creation can emerge from destruction—isn't just poetic. It's built into the physics of the universe. Energy is never lost; it's transformed. When a black hole releases radiation or jets, that energy doesn't vanish into nothing. It moves outward, changing the temperature, density, and chemistry of the gas it encounters. The result is a constant recycling of matter, turning what might seem like chaos into structure.

Even the material that escapes a black hole's grip plays a role in this cycle. As jets and winds travel outward, they enrich the surrounding space with heavier elements forged in earlier generations of stars. Those elements—carbon, oxygen, nitrogen, iron—become the building blocks for new stars, planets, and even life. In this way, the influence of black holes extends far beyond their immediate surroundings. They're participants in the cosmic chain that leads from gas and dust to galaxies, to solar systems, and eventually, to beings capable of wondering about them.

There's another layer to this story: the timing of these events. In many galaxies, periods of intense black hole activity coincide with bursts of star formation. The two phenomena rise and fall together, suggesting they're linked by more than coincidence. When a galaxy's core lights up as an active nucleus, the surrounding regions often become hotbeds of stellar birth. As the black hole settles down, the rate of star formation drops too. It's as if the galaxy and its central engine breathe in sync—one feeding the other in a grand, rhythmic exchange.

Modern simulations have helped reveal how this might happen in detail. When matter falls toward the black hole, magnetic fields and turbulence launch outflows that travel through the galaxy. The energy from those outflows compresses gas in some regions while dispersing it in others. Over time, this redistribution creates a self-regulating cycle. The black hole prevents its galaxy from forming too many stars at once but ensures that new stars continue to form steadily in the right places. Without that regulation, galaxies could burn out prematurely, exhausting their fuel long before the universe reached its current age.

In that sense, black holes act as cosmic moderators. They prevent extremes—too much creation or too much destruction. They bring balance to an otherwise unstable process, turning the violent energy of accretion into a force that sustains the long-term health of galaxies. It's an idea that challenges the old image of black holes as simple vacuums in space. They're more like engines of change, converting gravitational energy into the light and heat that shape everything around them.

9

COULD WE EVER VISIT ONE?

Trying to picture what it would feel like to fall into a black hole pushes human imagination to its limit. It's an experience no one could survive, yet it's fascinating to explore because it reveals how differently the universe behaves near such an extreme object. The forces at work are so far beyond anything we encounter in everyday life that words like "falling" or "descending" don't fully capture it. Still, the principles of physics allow us to describe what would happen, step by step, to anyone—or anything—unfortunate or curious enough to cross that final line.

At first, approaching a black hole wouldn't look especially dramatic. From a great distance, space behaves normally. Stars would still shine, and planets would still orbit as usual. Only when you began to close in would things start to change subtly. Gravity would become stronger in a way that's hard to feel but easy to measure. The closer you moved, the more you'd notice how light bent around the black hole. Stars behind it would smear into arcs or rings of distorted light, forming what's called a gravitational lens. It would be a hauntingly beautiful sight—an optical illusion caused not by glass or mirrors, but by the bending of space itself.

As you drew nearer, you'd see that the black hole wasn't entirely black. The material swirling around it—gas, dust, and debris pulled from nearby stars—would glow fiercely in the accretion disk. That disk would radiate energy across the spectrum, from visible light to high-energy X-rays, making the region around the black hole one of the brightest in the universe. The contrast would be stark: an invisible center surrounded by a ring of fire, light curving inwards toward a darkness no one could penetrate.

If you were somehow able to travel closer still, your surroundings would begin to behave in strange ways. Time, for one, would not flow evenly. According to Einstein's theory of relativity, gravity affects time. Near massive objects, time slows relative to places where gravity is weaker. As you approached the black hole, your clock would tick more and more slowly compared to someone far away. To you, everything would seem normal—you'd still feel seconds passing as usual. But to an observer watching from a safe distance, your movements would slow to a crawl. As you reached the edge—the event horizon—you would appear to freeze in place.

That illusion happens because light leaving your body would struggle to climb out of the black hole's gravity. Each signal would take longer to reach the outside world and would stretch into longer wavelengths. Eventually, your image would fade completely, swallowed by the blackness. From your perspective, though, nothing special would seem to happen as you crossed the horizon. You'd keep falling, unaware that the rest of the universe had effectively lost sight of you forever.

This is where the experience would shift from strange to fatal. The gravitational pull near a black hole isn't uniform —it's much stronger closer to the center. If you were falling feet-first, the gravity at your feet would be vastly greater than the gravity at your head. This difference creates what physicists call a tidal force. On Earth, tides are gentle, caused by the slight difference in the Moon's gravity between the near and far sides of our planet. Near a black hole, that difference becomes catastrophic. It would stretch you vertically while compressing you horizontally, pulling you into a long, thin line. This process, known colloquially as spaghettification, would happen in a fraction of a second.

The exact moment it occurs depends on the size of the black hole. In a smaller one, the tidal forces near the event horizon are so strong that you'd be torn apart long before crossing it. In a supermassive black hole, like the one at the center of our galaxy, the forces are milder at the horizon because the gradient of gravity changes more gradually. You could, in theory, pass through the horizon intact—but only briefly. The closer you fell toward the singularity at the center, the faster those tidal forces would grow, until they ripped apart every atom in your body.

It's tempting to think of that singularity as a point in space, but it's more accurate to call it a region where our laws of physics stop working. Einstein's equations predict infinite density and zero volume—a place where space and time curve infinitely. In reality, nature probably avoids true infinities, but we don't yet know how. Quantum physics, which governs the behavior of the smallest particles, might eventually explain what happens there. For now, it remains a mystery. Whatever it is, the singularity marks the end of any possible journey.

Some scientists have speculated that falling into a black

hole might not necessarily mean destruction, but transformation. The information carried by your particles—the arrangement of their energy and quantum states—might somehow survive, encoded on the surface of the event horizon or released later as radiation. This is part of what's known as the information paradox. It's an unresolved question in physics: does information truly vanish in a black hole, or is it preserved in some altered form? If the latter is true, then even in death, something of you would remain imprinted on the cosmos, woven into the fabric of spacetime itself.

To understand the experience from another perspective, think about how your surroundings would look as you fell. The light from stars behind you would appear to blue-shift, becoming brighter and more energetic, while light from ahead would red-shift, dimming as it stretched. Space itself would seem to funnel toward the center, like a tunnel narrowing endlessly. The black hole wouldn't appear as a hole at all but as a sphere of darkness surrounded by distorted light from all directions. Your sense of direction would become meaningless; gravity would be pulling equally on every side. Down would no longer exist in any normal sense.

As the singularity drew closer, the curvature of spacetime would become so extreme that paths of light—called geodesics—would converge. Every possible direction would lead inward. For all practical purposes, there would be no way to stop, turn around, or even signal outward. The laws of physics would leave you no choices. That inevitability, that one-way nature of the fall, is what makes black holes such powerful symbols of finality.

Yet, even in that inevitability, there's something strangely elegant. A black hole doesn't tear apart the universe; it simply enforces its rules to the fullest. Gravity, taken to its extreme, becomes an event horizon—a surface that marks the boundary between what can interact with the universe and what cannot. The fall into it is not chaotic but orderly, governed by predictable equations right up until the very end, where predictability itself dissolves.

If there were a way to observe the fall from within, it would challenge every intuition we have about space and time. You might see the entire history of the outside universe flash before you in an instant. As you accelerated inward, time outside would appear to speed up infinitely. Galaxies would form, age, and fade in a blink. The universe itself might run its full course while, for you, only moments passed. Whether that perception is meaningful or even possible depends on how consciousness interacts with time under those conditions—a question far beyond our current understanding.

No one will ever know firsthand what falling into a black hole feels like, but through mathematics and observation, we can trace its outlines. We can say how gravity stretches, how time distorts, how light bends. We can describe the path of matter as it falls, and we can estimate the energies involved. Yet the experience itself—the sensory, subjective reality of it—remains unreachable. It exists at the border between science and imagination, a scenario that defines the limits of both.

THE LIMITS OF OUR CURRENT SPACECRAFT AND TECHNOLOGY

The dream of exploring a black hole faces an unavoidable obstacle: our technology simply isn't built for the scale of the universe. Even our most advanced spacecraft—those that have pushed the boundaries of exploration—barely skim the surface of what's possible when compared to the distances, forces, and energies involved in reaching something like a black hole. Understanding these limits helps make clear not only how far we've come, but how far we still have to go before interstellar travel becomes more than theory.

To start with perspective, the farthest human-made object from Earth is *Voyager 1*. Launched in 1977, it's been traveling through space for nearly half a century and is now over 15 billion miles away. That sounds immense—and it is—but in cosmic terms, it's only about 0.002 light-years. The nearest star, Proxima Centauri, is 4.24 light-years away. A black hole, even the closest known one, lies thousands of light-years beyond that. At Voyager's speed, it would take tens of thousands of years just to reach the nearest star, and millions to reach a black hole. That comparison alone shows how vast the gulf is between what we can do and what the universe allows.

Our current propulsion systems are the main limitation. Rockets today rely on chemical propulsion—the same principle used since the earliest space programs. They burn fuel to produce thrust, pushing against the emptiness of space by ejecting hot gases backward. This method works well for getting off Earth and maneuvering in orbit, but it's incredibly inefficient for interstellar distances. Most of a rocket's weight is fuel, and even the most powerful engines barely scratch a fraction of light speed. A journey to a black hole would require a propulsion system capable of maintaining high speeds for decades or even centuries without refueling, something far beyond what chemical rockets can manage.

There have been many proposals to overcome that problem. Nuclear propulsion, for example, could generate far more energy from much less fuel, potentially increasing travel speed by several orders of magnitude. Some designs involve fusing atoms, similar to what powers the Sun, to produce controlled bursts of thrust. Others suggest using antimatter—the exact opposite of normal matter—which releases enormous energy when it comes into contact with regular atoms. These ideas exist mostly on paper because building and containing such reactions safely would require technology that doesn't yet exist. Even if it did, protecting a spacecraft and its crew from the radiation produced by these engines would be a monumental engineering challenge.

Another concept involves using light itself as propulsion. A spacecraft equipped with a massive reflective sail could be pushed by lasers or sunlight, slowly accelerating over long distances. This idea, called a *light sail*, has already been tested on a small scale. It's efficient in theory because it doesn't require carrying fuel, but it comes with practical issues. To reach significant speeds, the sail would need to be huge—possibly kilometers wide—and precisely aimed by powerful ground-based lasers. Any slight misalignment could send the craft spinning off course. Even at its best, a light sail would take decades just to reach the nearest star, and a black hole would remain far beyond its reach.

Then there's the problem of communication. Even if a probe could travel to a black hole, sending information back would be nearly impossible. Radio signals, which move at the speed of light, would take thousands of years to reach Earth. And the farther away the spacecraft is, the weaker its signal becomes. Our deep-space antennas, like those used to contact Voyager, are already operating near the edge of their sensitivity. Extending that to black hole distances would require a massive upgrade in power, precision, and data-handling technology.

Durability poses another limitation. Spacecraft that travel for decades must survive extreme conditions: cosmic radiation, micrometeoroids, and the temperature extremes of deep space. Even the smallest puncture or electronics failure could end a mission. Voyager 1 and 2 are testaments to human engineering, but they're already struggling as their power sources weaken. A journey lasting centuries or millennia would demand machines capable of self-repair, possibly even artificial intelligence to manage evolving problems without human help. That's a level of autonomy we haven't achieved yet, especially for something operating so far from home.

Human travel is even more complex. Keeping people alive in space requires not just propulsion and power, but food, water, oxygen, and protection from radiation. The International Space Station, orbiting a mere 250 miles above Earth, needs constant resupply and maintenance. A mission beyond the Solar System would need closed-loop life support systems capable of recycling every drop of water and breath of air. That technology is in development, but nowhere near perfect. Radiation from cosmic rays and solar flares could cause long-term health damage or death without substantial shielding—shielding that would make the spacecraft heavier, requiring even more energy to move. Every solution seems to create new problems.

There's also the question of time. Even if we developed a ship capable of traveling at a tenth the speed of light—an incredible accomplishment—it would still take tens of thousands of years to reach a black hole. Generational ships, where entire societies live and die during the voyage, have been proposed, but that introduces social and ethical questions we haven't begun to solve. How would future generations maintain the mission's purpose? What happens if their descendants decide they no longer wish to continue? These are challenges of human endurance as much as technological capability.

In some ways, the problem isn't just that we lack the technology—it's that the physics itself sets hard limits. The speed of light is the ultimate barrier. According to Einstein's theory of relativity, nothing with mass can reach or exceed it. As an object approaches light speed, its energy requirement grows without bound. Every step closer takes exponentially more energy. That means no matter how advanced we become, there will always be destinations that are effectively unreachable within a human lifetime.

Still, scientists continue to explore ways to bend those limits. Theoretical ideas like the Alcubierre "warp drive" propose compressing space in front of a ship and expanding it behind, effectively moving the craft by distorting spacetime itself. In that scenario, the ship doesn't technically move faster than light—it rides a wave of curved space. But the energy needed to create such a distortion would be immense, possibly greater than the total energy of entire stars. The concept remains mathematically intriguing but physically out of reach for now.

Despite these limitations, progress is real and steady. The technologies we take for granted today—satellite communication, GPS, robotic rovers—were once unimaginable. Breakthroughs in materials science, artificial intelligence, and quantum computing could open paths we can't yet foresee. Miniaturized electronics, for example, allow spacecraft to be lighter and more efficient than ever before. Advances in propulsion research, like ion thrusters and plasma drives, show promise for long-term missions within the Solar System. Each small step brings us closer to expanding what's possible, even if the timeline is measured in centuries rather than decades.

TIME TRAVEL, WORMHOLES, AND SCIENCE FICTION VS. SCIENCE FACT

Few ideas capture the imagination quite like time travel. The ability to step outside the flow of time—to revisit the past or leap ahead into the future—has fueled countless stories, movies, and theories. It's both thrilling and unsettling because time governs everything we do, and the idea of escaping it touches something deep in the human psyche. When black holes enter that conversation, the line between science fiction and science fact begins to blur in fascinating ways.

At the heart of it all lies Einstein's theory of relativity, which fundamentally changed how we understand time and space. Before Einstein, time was considered absolute—it ticked at the same rate everywhere, independent of motion or gravity. Relativity shattered that idea. It showed that time is woven together with space into a single fabric called spacetime. The speed at which time passes depends on how fast you move and how strong gravity is where you are. That means time isn't fixed—it can stretch, slow, or even appear to stop relative to other points in the universe.

In that sense, time travel—at least in one direction—is already real. Every astronaut who's ever orbited Earth has technically experienced it. When they move at high speeds in weaker gravity, their clocks tick slightly slower than those on the ground. The difference is tiny, fractions of a second, but measurable. The effect becomes more dramatic as speeds approach the speed of light or as gravity increases near massive objects like black holes. The closer you get to those extremes, the greater the distortion of time becomes.

This phenomenon, known as time dilation, means that a traveler could leave Earth, circle a black hole at high speed, and return to find that years—or even centuries—had passed on Earth while only a short time had passed for them. It's not the kind of time travel popularized in movies, where people can jump freely between eras, but it's still a kind of travel through time. The universe itself allows it; it's baked into the equations of physics.

The more complicated question is whether we could ever move backward in time. That's where the concept of wormholes comes in. A wormhole is a hypothetical tunnel connecting two points in spacetime. If such a tunnel existed, it could, in theory, allow instant travel between distant regions—or even different moments—of the universe. You can think of spacetime as a sheet of paper. If you fold that paper in half and poke a hole through it, you've created a shortcut between two distant points. That's the basic idea of a wormhole: a bridge that bends space and time together to make travel faster than light without breaking the laws of relativity.

Wormholes weren't dreamed up by science fiction writers. They emerged naturally from Einstein's equations, discovered as possible—but unstable—solutions to how gravity warps spacetime. In theory, a black hole and a white hole (its hypothetical opposite, where matter and energy are expelled rather than absorbed) could be connected to form a wormhole. The problem is that such a structure would collapse almost instantly. The gravity pulling inward would crush it before anything could pass through. To keep it open, something would need to counteract that pull—something with "negative energy" or "exotic matter."

Negative energy isn't something we've ever observed directly, but quantum physics suggests it might exist in tiny amounts under special conditions. If it could be produced and controlled, it might stabilize a wormhole long enough for information—or maybe even a person—to travel through. But this is where we drift from established physics into speculation. The equations allow the idea, but the universe may not cooperate in practice. Even if a wormhole could be created, keeping it open safely might require more energy than the total output of entire galaxies.

Still, scientists study wormholes seriously because they challenge the boundaries of what's possible. They represent the intersection of relativity and quantum mechanics, two of the most powerful but still incomplete theories in modern science. If wormholes exist naturally—perhaps on microscopic scales—they might help explain how the universe is connected at its deepest level. Some theories even suggest that black holes themselves could hide wormholes within them, linking distant regions of space. There's no evidence for this yet, but it's one of the many mysteries that keeps astrophysicists working late into the night.

The idea of traveling backward in time, though, comes with more than just technical problems—it runs headfirst into paradoxes. If you could return to the past, you might affect events that would prevent your own trip from happening in the first place. The classic example is the "grandfather paradox": if you traveled back and somehow stopped your grandparents from meeting, you'd erase your own existence—and therefore your ability to go back and change anything. Physics tends to avoid such contradictions. Some interpretations of relativity suggest that if time loops were possible, events would adjust themselves in ways that preserve consistency. You might go back, but you couldn't change anything that would alter your own timeline.

Other ideas propose a multiverse, where each action branches into a new universe. In that framework, traveling to the past wouldn't change your own timeline—it would create or move you into a different one. While this kind of thinking makes for great storytelling, there's no experimental evidence that such branching universes exist. They remain intriguing but speculative interpretations of quantum theory.

Science fiction has long blurred these ideas with narrative imagination, often stretching physics beyond its breaking point. Wormholes in movies like *Interstellar* or *Contact* serve as elegant storytelling devices—ways to connect distant places and moments without waiting thousands of years. Time machines in stories let us explore moral and philosophical questions about cause and effect, destiny, and free will. The best science fiction doesn't just imagine new technologies; it tests human values in the face of impossible choices. What would you change if you could? What would you risk to see the future?

Science, on the other hand, works within stricter boundaries. It doesn't deal in "what if" so much as "what can be." And within those limits, time travel remains largely one-directional. Time can be slowed by gravity or speed, but it can't yet be reversed. Wormholes may exist in theory, but they're unstable, fleeting, and unproven. For now, the universe seems to guard its chronology carefully.

Yet, there's something telling about how persistent these ideas are. Every time science advances, fiction races to meet it, and sometimes the two trade places. The notion of black holes themselves was once considered science fiction—a strange mathematical oddity that couldn't possibly exist.

Now we've photographed one. What was fantasy a century ago is now confirmed reality. It's not unreasonable to think that some of today's "impossible" ideas might someday move into the realm of the possible, though perhaps not in the way we expect.

Even if we never build a machine to leap through time, we already live in a universe where time behaves differently than intuition suggests. The GPS in your phone corrects for time dilation every second—without those adjustments, navigation would drift miles off course in minutes. Astronauts aboard the International Space Station experience time slightly slower than people on Earth's surface. These aren't speculative effects; they're measurable, practical outcomes of Einstein's relativity. We experience time travel, in a sense, every day, though it's too subtle to notice.

10

THE FUTURE OF BLACK HOLE RESEARCH

Every generation of astronomers inherits a bigger window to the universe. A few centuries ago, telescopes showed us the moons of Jupiter and the rings of Saturn. In the last few decades, we've peered into distant galaxies and captured the first blurry outlines of black holes themselves. But the next generation of telescopes and space missions will do more than just expand what we can see—they'll let us explore how the universe behaves under conditions that push physics to its limits.

The story of black hole research has always been tied to the evolution of technology. When Einstein first wrote down his equations of general relativity, telescopes weren't powerful enough to test their predictions. The first real clues came from radio astronomy, when scientists found mysterious signals from compact objects too massive to be stars. Those turned out to be neutron stars and black holes. Now, as instruments grow sharper and more sensitive, the questions are shifting from *if* black holes exist to *how* they form, evolve, and shape the universe.

At the center of this new era are telescopes built to

capture light that human eyes can't see. Black holes themselves don't emit visible light, but the gas and dust swirling around them radiate intensely in other parts of the spectrum—especially radio waves, X-rays, and gamma rays. Studying these wavelengths gives astronomers a way to trace what's happening in the extreme environments near event horizons.

The Event Horizon Telescope (EHT), which linked radio observatories around the globe to create a planet-sized virtual telescope, was just the beginning. Its first image of the black hole in galaxy M87 stunned the world, showing a glowing ring of light around a dark shadow. Now, an expanded version of the project is being planned. Future upgrades will add new observatories and better synchronization systems, sharpening the image resolution by factors of two or more. Scientists hope to capture time-lapse sequences—essentially "movies" of the material orbiting black holes—to watch their behavior in real time.

Meanwhile, other instruments will extend the reach of these observations. The *Next Generation Very Large Array* (ngVLA), planned for the 2030s, will be a network of hundreds of radio antennas spread across North America. It will have the sensitivity to study the magnetic fields around black holes, helping to answer how they launch powerful jets of energy that stretch across galaxies. Those jets are among the most mysterious features in astrophysics—they carry more energy than anything else in the universe, yet their exact mechanism remains unclear.

For X-ray observations, a new wave of space telescopes will peer even closer to the energetic edges of black holes. Missions like *Athena* (Advanced Telescope for High-Energy Astrophysics), being developed by the European Space Agency, aim to measure how hot gas moves and glows as it falls inward. By tracking how that light shifts in energy, scientists can map the motion of matter near the event horizon and determine how fast a black hole spins. Spin is more than a statistic—it affects how the black hole interacts with its surroundings, how it warps spacetime, and how much energy it can release.

In addition to light, the next frontier of exploration will rely on gravity itself as a messenger. Ground-based observatories like LIGO and Virgo have already detected gravitational waves from colliding black holes. But a new generation of detectors is on the way, expanding both range and sensitivity. The *Einstein Telescope* in Europe and *Cosmic Explorer* in the United States will be massive underground facilities capable of detecting black hole mergers from the earliest epochs of the universe. These instruments will pick up faint ripples from billions of light-years away, allowing scientists to trace how black holes have evolved over cosmic time.

Perhaps the most ambitious of all these projects will orbit the Sun rather than sit on Earth. The *Laser Interferometer Space Antenna* (LISA), scheduled for launch in the 2030s, will consist of three spacecraft flying in a triangle millions of kilometers apart. Using laser beams to measure minute distortions in spacetime, it will detect low-frequency gravitational waves—signals produced by the mergers of supermassive black holes, or by intermediate ones that bridge the size gap between stellar and galactic scales. LISA could even pick up the faint hum of gravitational waves left over from the earliest moments after the Big Bang.

Other missions will look for the subtle fingerprints of black holes in the way they influence their surroundings. The *James Webb Space Telescope* has already begun spotting the light from some of the universe's first galaxies, and many of those early structures appear to host massive black holes at their centers. By studying how these ancient systems grew, Webb and its successors could uncover how black holes first formed—a puzzle that has lingered for decades. Were they born from the collapse of the first stars, or did they arise directly from enormous clouds of gas? Future infrared telescopes will continue probing that question by peering even farther back in time.

Not all new tools will be space-based. On the ground, advances in adaptive optics—technology that corrects for the blurring effects of Earth's atmosphere—are making it possible to observe stars orbiting close to the black hole at the center of our own Milky Way. The upcoming *Extremely Large Telescope* (ELT) in Chile will feature a mirror nearly 40 meters across, giving it unmatched precision. Watching those stars move, astronomers can measure how space bends under the influence of gravity near Sagittarius A*, our galaxy's central black hole. That data may provide some of the strongest tests yet of Einstein's theory of relativity.

Complementing these optical and radio giants will be smaller, specialized instruments focusing on particular wavelengths or phenomena. For example, the *Square Kilometre Array* (SKA), an enormous radio telescope project spanning parts of South Africa and Australia, will survey the sky in unprecedented detail. Its sensitivity could reveal entire populations of previously undetected black holes, including the elusive intermediate-mass ones that have long been theorized but rarely observed. These discoveries will fill in gaps in our understanding of how black holes bridge the size spectrum between collapsed stars and the giants at galaxy centers.

Even the quiet signals that black holes leave behind after they've finished consuming matter can be studied.

When a black hole stops actively feeding, it may appear "dormant," but its gravity still affects nearby stars and gas. Upcoming all-sky surveys, such as those planned by the *Vera C. Rubin Observatory*, will monitor the entire visible sky night after night. Over time, its vast dataset could reveal subtle motions or flickers caused by hidden black holes passing through stellar neighborhoods. For the first time, we may be able to build a true census of black holes across our galaxy.

Another intriguing area of research will focus on the smallest and most ancient black holes—those that might have formed soon after the Big Bang. These so-called "primordial black holes" are still hypothetical, but their existence could explain some mysteries, such as the nature of dark matter or the seeds of cosmic structure. Upcoming missions sensitive to gravitational lensing—the bending of light caused by massive objects—could detect their presence indirectly. If even a fraction of dark matter consists of such black holes, it would completely reshape our understanding of the universe's composition.

Beyond the telescopes and detectors themselves, a quiet revolution is happening in how scientists handle data. Modern observatories produce petabytes of information, far more than any single researcher can analyze. Artificial intelligence systems are becoming crucial partners in this process. They can spot patterns in the noise of telescope signals, identify potential black hole candidates, and even predict where to look next. Instead of replacing astronomers, these tools amplify their ability to interpret the universe, uncovering hidden relationships that might otherwise go unnoticed.

EPILOGUE – WONDER AT THE EDGE OF SPACE

There's a peculiar pull that certain mysteries have over us—the ones we know we might never fully solve. Black holes are one of those mysteries. Even people who have no interest in physics or astronomy tend to pause when they hear about them. There's something magnetic about the idea of a place in space where light itself can't escape, where the normal rules of reality seem to stop working. It's as if black holes sit right at the border between knowledge and wonder, and we can't resist standing there, looking out into the dark.

Part of the draw comes from what they represent. A black hole is a contradiction made real—a thing that exists by erasing everything that falls into it. It's both a presence and an absence, a proof that the universe is far stranger than common sense allows. We're drawn to contradictions because they challenge us. They remind us that the world doesn't fit neatly into the boxes we build for it. In a way, studying black holes isn't just about understanding the universe—it's about learning to live comfortably with uncertainty.

Human beings have always been explorers, but not just in the physical sense. We explore ideas, questions, and the edges of comprehension. The same instinct that once drove people to cross oceans now drives scientists to study the invisible. A telescope pointed at a galaxy billions of light-years away is no less an act of adventure than a ship setting out for an unknown shore. Black holes, being the ultimate unknowns, are irresistible to that kind of curiosity. They represent both danger and discovery—the possibility of being lost and the promise of finding something extraordinary.

There's another reason they fascinate us: they reflect back our own limits. Every time we hit a boundary of knowledge—every time the math fails or a theory falls apart—it's like standing before a cosmic mirror. We see how much we've learned, but also how much remains outside our reach. That gap between what we know and what we can't yet explain is where curiosity thrives. It's a space that invites humility without discouragement. Black holes sit right in that gap, constantly reminding us that even the smartest minds and the most advanced tools can't answer everything. And yet, we keep trying.

It's no coincidence that some of the greatest leaps in physics have come from questions about black holes. They've pushed scientists to rethink space, time, energy, and even the nature of reality itself. But they've also captured the public imagination in ways few scientific topics ever do. Everyone understands what it means to face something you can't quite grasp. There's a universal resonance in the idea of staring into the unknown and being both frightened and inspired by it.

Mystery, after all, is part of what gives life its texture. If everything were fully explained, if every question had a clean answer, the world would lose a kind of depth. Black holes remind us that the universe is layered—that there are truths waiting just beyond what we can see or measure. And in that mystery, there's meaning. Not necessarily in the answers we find, but in the search itself. The act of wondering connects us to something larger than ourselves.

For many people, black holes spark both scientific and philosophical reflection. They make us ask questions that go beyond equations: What is time, really? What happens when something ends? Can something truly disappear, or does it always leave a trace? These aren't just astrophysical questions—they're human ones. When we think about a star collapsing into a black hole, we're reminded that endings and beginnings are often intertwined. Death becomes transformation; disappearance becomes another form of existence. There's comfort in that symmetry.

At the same time, black holes remind us of scale. They put everything into perspective. The size of one, the forces it wields, the distances involved—they make even the grandest human achievements seem small. But that's not a reason to feel insignificant. Quite the opposite. The fact that tiny creatures on a small planet orbiting an average star can even *know* about black holes is astonishing. We're part of a universe vast enough to contain things that defy our understanding, yet we've evolved brains capable of detecting and describing them. It's a beautiful irony: the cosmos produces beings capable of reflecting on the cosmos itself.

Mysteries like black holes also connect generations. Every discovery builds on the work of those who came before, and every unanswered question becomes a challenge for those who come next. When scientists study them, they're contributing to a conversation that began long before they were born and will continue long after they're gone.

That continuity—the passing of curiosity from one mind to another—might be the closest thing we have to immortality. Even if the universe is destined to cool, fade, or collapse, the search for understanding leaves a mark.

There's something poetic about how much we learn by studying things we can never touch. Black holes teach us about the birth of the universe, the behavior of matter, and the shape of time, all from a distance. It's a reminder that discovery doesn't always require direct contact—it can come from patience, observation, and a willingness to question everything we think we know. In that sense, black holes symbolize one of humanity's finest qualities: our refusal to stop asking questions, even when the answers seem unreachable.

They also speak to our emotional lives in subtle ways. The pull of a black hole, the way it draws everything inward, mirrors our own gravitational pull toward meaning. We can't help but be drawn to what we don't fully understand. Whether it's love, art, death, or the cosmos, we're wired to seek connection with the mysterious. It's not just about solving problems—it's about feeling part of something greater, even if that something remains partly hidden.

Black holes stand as reminders that mystery doesn't mean chaos or randomness. Even their strangeness follows precise laws, even if those laws sometimes baffle us. There's an order in the unknown, waiting to be uncovered. That balance between mystery and structure mirrors life itself. We live with questions—about purpose, destiny, and existence—but within those questions, there's still rhythm and beauty.

As we continue to explore, black holes will keep challenging us to think bigger, not only about the universe but about our place within it. They remind us that knowledge isn't a finish line—it's a horizon that moves as we approach it. Every new discovery opens up new territory to explore. That endlessness is part of what makes science, and life, endlessly compelling.

When you strip away the technical language and equations, what's left at the heart of black hole research is wonder. A kind of curiosity that blends logic with imagination, precision with awe. It's the same feeling that led ancient stargazers to trace patterns in the night sky and ask what lies beyond. The tools have changed, but the impulse is timeless. To be human is to be curious—to stand beneath a vast, mysterious sky and reach for understanding even when we know we might never have all the answers.

QUICK REFERENCE GLOSSARY

EVENT HORIZON

The invisible edge around a black hole where gravity becomes so strong that nothing—not even light—can escape.

Singularity

A point at the center of a black hole where matter is crushed to infinite density and our known laws of physics no longer work.

Accretion Disk

A glowing ring of superheated gas and dust spiraling around a black hole before falling in.

Relativity

Einstein's idea that space and time are linked, and that gravity is the result of mass bending the fabric of both.

Space-Time

The four-dimensional "fabric" that combines space and time into one continuous whole.

Gravitational Wave

A ripple in space-time created when massive objects like black holes move or collide.

Redshift

The stretching of light into longer wavelengths as it moves away from us, showing the universe's expansion.

Blue Shift

The shortening of light waves as an object moves closer, making it appear bluer and more energetic.

Photon

A tiny packet of light that carries energy but has no mass.

Quantum Mechanics

The science that describes how matter and energy behave on the smallest possible scales.

General Relativity

Einstein's theory explaining how gravity works through the warping of space-time, not by invisible "pulls."

Special Relativity

Einstein's theory showing that time and space are flexible, depending on how fast something moves.

Gravitational Lensing

The bending of light around massive objects, letting us see things hidden behind them.

Hawking Radiation

Faint heat emitted from the edge of a black hole, meaning black holes can slowly lose energy over time.

Evaporation (of Black Holes)

The gradual shrinking and disappearance of a black hole as it loses energy through Hawking radiation.

Wormhole

A theoretical tunnel that could connect two distant points in space-time, creating a shortcut through the universe.

Tidal Forces

The stretching and squeezing caused by gravity acting unevenly on an object in a strong gravitational field.

Spaghettification

What happens when those tidal forces stretch matter into a long, thin line as it falls toward a black hole.

Supermassive Black Hole

A black hole millions or billions of times heavier than our Sun, usually found at the center of galaxies.

Stellar Black Hole

A smaller black hole formed when a massive star collapses after running out of fuel.

Intermediate Black Hole

A mid-sized black hole, thought to bridge the gap between stellar and supermassive ones, though few have been confirmed.

Primordial Black Hole

A tiny black hole that may have formed in the early universe, possibly as small as an atom but incredibly dense.

Sagittarius A*

The supermassive black hole sitting quietly at the heart of our Milky Way galaxy.

M87*

The massive black hole in the galaxy Messier 87, the first one ever photographed by the Event Horizon Telescope.

Gravitational Pull

The attraction that all matter exerts on other matter, strongest in the most massive objects.

Curvature of Space

The bending of the universe's geometry caused by the presence of mass and energy.

Time Dilation

The slowing of time near strong gravity or at high speeds, where clocks run differently depending on perspective.

Photon Sphere

The zone around a black hole where light orbits in perfect circles before escaping or being pulled in.

Jet (or Relativistic Jet)

A stream of particles and radiation blasted out from near a black hole at nearly the speed of light.

Magnetic Field

An invisible force created by moving charged particles, guiding the behavior of matter around stars and black holes.

Cosmic Microwave Background

The faint leftover glow from the Big Bang, filling all of space with ancient light.

Dark Matter

The invisible mass that holds galaxies together through gravity, even though it emits no light.

Dark Energy

A mysterious pressure that causes the universe's expansion to speed up instead of slowing down.

The Big Bang

The beginning of the universe, when space, time, and matter first expanded from an unimaginably hot and dense state.

Entropy

A measure of disorder or randomness; in physics, it tends to increase over time, driving the universe toward balance.

Information Paradox

The unresolved question of whether the data about matter falling into a black hole is lost forever or somehow preserved.

Quantum Fluctuation

Tiny, random changes in energy that constantly occur in empty space.

Cosmic Horizon

The limit of what we can see in the universe—the farthest distance light has reached us since the Big Bang.

Escape Velocity

The speed an object must reach to break free from another object's gravity.

Light-Year

The distance light travels in one year—about six trillion miles, or nine trillion kilometers.

Parsec

A larger unit of astronomical distance, equal to about 3.26 light-years.

Orbit

The curved path an object follows as it moves around a more massive body due to gravity.

Mass

The amount of matter in something, which determines how much gravity it produces.

Density

How much mass is packed into a given volume; black holes have the highest density possible.

Energy

The capacity to do work or cause change, found in everything from light to motion to gravity itself.

Momentum

The tendency of an object in motion to keep moving in the same direction unless acted on by another force.

Fusion

The process by which stars produce energy, merging hydrogen atoms into helium and releasing vast amounts of light.

Supernova

A massive explosion that marks the death of a star, often leaving behind a neutron star or black hole.

Neutron Star

The collapsed core of a massive star, made almost entirely of neutrons and so dense that a teaspoon would weigh billions of tons.

Pulsar

A spinning neutron star that emits beams of radiation, appearing to pulse as it rotates.

Galaxy

A vast system of stars, gas, dust, and dark matter held together by gravity.

Milky Way

Our home galaxy, a spiral collection of hundreds of billions of stars—and one supermassive black hole at its center.

Cluster (of Galaxies)

A large group of galaxies bound together by gravity, often containing thousands of members.

Quasar

An extremely bright, energetic object powered by matter falling into a supermassive black hole.

Red Giant

A stage in a star's life when it expands and cools after using up most of its hydrogen fuel.

White Dwarf

A small, dense remnant of a star that has exhausted its fuel and collapsed inward.

Cosmic Inflation

A rapid expansion of the universe in its earliest moments, smoothing out its structure.

Multiverse

The speculative idea that our universe is just one of many, each with different properties or laws of physics.

Relativistic Speed

A speed so close to the speed of light that relativistic effects—like time dilation and mass increase—become significant.

Planck Scale

The smallest possible size in physics, where space and time themselves may become "quantized."

Singular Point

Another way to describe a singularity—a point where known physics breaks down.

Magnetar

A highly magnetic type of neutron star capable of releasing enormous bursts of energy.

Telescope Array

Multiple telescopes working together to observe the same object, increasing resolution and detail.

Interstellar Medium

The gas, dust, and plasma that fill the space between stars within a galaxy.

Relativistic Effect

Any change in perception or measurement that occurs because of the effects of relativity, such as time or distance distortion.

Cosmic Ray

High-energy particles, mostly protons, traveling through space at nearly the speed of light.

Thermodynamics

The study of how heat, energy, and work interact and transform.

Plasma

A state of matter consisting of charged particles—often found in stars and around black holes.

Rotation Curve

A graph showing how fast stars move in a galaxy, used to detect dark matter.

Energy Conservation

The principle that energy cannot be created or destroyed, only transformed from one form to another.

Curved Light Path

The way light bends when it travels near a massive object, proving that gravity shapes space itself.

Blackbody Radiation

The heat emitted by an object that absorbs all light falling on it—used to study stars and cosmic backgrounds.

www.ingramcontent.com/pod-product-compliance
Lightning Source LLC
LaVergne TN
LVHW090401210926
848480LV00002B/419